BASIC ELECTRONICS C

3.

621.381 PLA

**LEARNING.**
••••••••••••••services

**Cornwall College Camborne**
*Learning Centre - FE*

01209 616259

This resource is to be returned on or before the last date
stamped below. To renew items please contact the Centre

## Three Week Loan

I0418296

# Hodder & Stoughton

## LONDON SYDNEY AUCKLAND TORONTO

A moment's insight is sometimes worth a life's experience.

*The Professor at the Breakfast Table*
Oliver Wendell Holmes, Snr.

TECHNOLOGY WORKSHOP
CORNWALL COLLEGE
POOL, REDRUTH,
CORNWALL TR15 3RD

Every effort has been made to trace and acknowledge ownership of
copyright. The publishers will be glad to make suitable
arrangements with any copyright holders whom it has not been
possible to contact.

© 1990 SCDC Publications

First published in Great Britain 1976

Second edition 1990

*British Library Cataloguing in Publication Data.*
Plant, M. (Malcolm), 1936–
  Basic electronics. — 2nd ed
  Book C. Diodes and transistors
  1. Electronic equipment. For schools
  I. Title
  621.381
  ISBN   0 340 41493 6
  ISBN   0 340 41490 1 set

All rights reserved. No part of this publication may be reproduced
or transmitted in any form or by any means, electronic or
mechanical, including photocopy, recording, or any information
storage and retrieval system, without permission in writing from
the publisher or under licence from the Copyright Licensing
Agency Limited. Further details of such licences (for reprographic
reproduction) may be obtained from the Copyright Licensing
Agency Limited, of 33–34 Alfred Place, London WC1E 7DP.

Typeset in Baskerville by Taurus Graphics, Abingdon, Oxon.
Printed for the educational publishing division of Hodder and
Stoughton Ltd., Mill Road, Dunton Green, Sevenoaks, Kent, TN13
2YA by Thomson Litho Ltd, East Kilbride.

Basic Electronics is published in five parts

**Book A** Introducing Electronics
ISBN   0 340 41495 2

**Book B** Resistors, Capacitors and Inductors
ISBN   0 340 41494 4

**Book C** Diodes and Transistors
ISBN   0 340 41493 6

**Book D** Analogue Systems
ISBN   0 340 41492 8

**Book E** Digital Systems
ISBN   0 340 41491 X

It is also available as one complete volume:
ISBN   0 340 41490 1

## Note about the author

Malcolm Plant is a Principal Lecturer in the Faculty of Education at Nottingham Polytechnic. He is the author of several books, including *Teach Yourself Electronics* (Hodder & Stoughton 1988). His main professional interests are in astronomy and astrophysics, electronics instrumentation and issues relating to conservation and the environment.

# Contents

## Diodes

# Transistors

# Summary

This book is largely about the properties and uses of diodes and transistors. Chapters 1 and 2, show that the diode has the property of passing current easily one way and preventing current flow in the opposite direction. This property makes diodes especially useful for changing alternating current into direct current — see Chapters 4, 5 and 6.

A form of rectification is used to detect amplitude modulated (AM) signals in radio receivers, and this principle is explained in Chapter 7. There are many special types of rectifier diodes, such as the Zener diode introduced in Chapter 8. This device is used to provide stabilised voltages in circuits such as power supplies — Chapter 9. Some diodes are sensitive to light (the photodiode), while others emit coloured light (the light emitting diode) as explained in Chapter 11.

Diodes are the simplest type of semiconductor device, and are made from the materials silicon or germanium. The electrical properties of these materials are explained in Chapters 12 and 13. But silicon and germanium are treated with 'impurities' to make them have the special properties required in semiconductor devices such as the diode. This process is called 'doping' which yields n-type and p-type semiconductors as explained in Chapter 14. Diodes are made by combining n-type and p-type materials to provide a p-n junction as described in Chapter 16.

The transistor is the most widely used semiconductor device; it is used either as a discrete (i.e. separate) device, or as one of many others integrated with other components on a silicon chip. There are two main types of transistor, the common bipolar transistor which is introduced in Chapter 17, and the unipolar, or field-effect transistor, which is described in Chapter 27.

The bipolar transistor has three terminals. The current flowing between the collector and emitter terminals is controlled by a much smaller current flowing into or out of the base terminal. This property enables transistors to be used as 'solid-state' switches as explained in Chapter 18. Chapter 20 describes how two transistors in the form of the Darlington pair provides very sensitive switching action. An introduction to some applications for the single and Darlington pair transistor switches is given in Chapter 21.

For those readers who are interested, Chapters 22 to 24 explain some of the characteristics of bipolar transistors which determine how they are used as switches and amplifiers. Some introductory ideas of positive and negative feedback are explained in Chapter 25. Negative feedback makes the performance of amplifiers more stable and less dependent on the characteristics of the transistors used. (This property of negative feedback is discussed again in Book D in relation to operational amplifiers.) Positive feedback enables transistor oscillators to be designed, e.g. the two-transistor astable which is explained in Chapter 25.

Chapter 26 explains the basic properties and applications of the field-effect transistor (FET). Unlike the bipolar transistor, the FET controls current flowing between its drain and source terminals by a voltage, not a current, applied to its gate terminal. This property provides some unusual applications for FETs, touch switches and timer circuits, for example. Chapter 27 introduces thyristors and triacs, while Chapter 28 provides a range of further simple applications for bipolar and field-effect transistors.

Chapter 29 contains a description of seven Project Modules. These are units for the rapid assembly of electronic systems. They can be bought from the supplier or built from the printed circuit board (PCB) design provided. You should turn to Chapter 12 in Book A for an introduction to their use.

As with every book of *Basic Electronics*, Book C ends with Answers to Questions in the text, Revision Questions and Revision Answers.

If you want to follow an easier and quicker route through Book C, you should omit the sections of the text marked with the symbols $\nabla$ and $\triangle$ in the left hand margin.

# 1 Diodes—An Introduction

## 1.1 **Reminder**

Resistors, capacitors, and inductors are three electronic components which you have already met. They are used in circuits to control the flow of electrons, but the way by which they exert this control is different for each component.

### *Remember*

The *resistor* will allow both alternating current (a.c.) and direct current (d.c.) to pass through it — see figure 1.1.

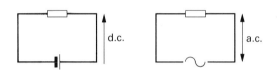

**Figure 1.1** Resistors pass a.c. and d.c.

The *capacitor* will allow only alternating current to flow through it. In a d.c. circuit, current will stop flowing through it when the capacitor is charged — see figure 1.2.

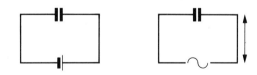

**Figure 1.2** Capacitors pass a.c. but not d.c.

The *inductor* will allow both direct and alternating current to flow through it, but it resists the flow of alternating current more than that of direct current — see figure 1.3.

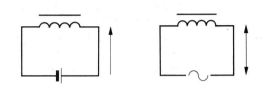

**Figure 1.3** Inductors pass a.c. less easily than d.c.

## *Questions*

1   What do you have to remember about the way to connect an electrolytic capacitor in a circuit?
2   How does the resistance (or reactance) of an inductor vary as the frequency of an alternating current increases?

## 1.2 **The purpose of a diode**

The diode is an electronic component which allows current to flow through it easily in one direction but opposes the current flow in the opposite direction. As you will see, this *rectifying* action of a diode is put to good use in radio receivers and power supplies. The electronic symbol for the diode is:

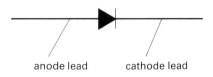

Conventional current flows in the direction of the arrow. The diode has two leads known as the *anode* and the *cathode*.

The cathode lead is usually marked with a black, white or red band, as shown in figure 1.4(a) and (b). The cathode is the terminal from which conventional current flows when the diode is carrying current in the easy direction. Thus, conventional current flows from left to right in the examples shown. On some diodes the symbol is marked, so that there is no confusion as to which is the cathode and which the anode lead.

*Germanium* (Ge) and *silicon* (Si) are two very important *semiconducting materials*. They are used to manufacture diodes, transistors, and other solid-state devices. More is said about these semiconducting materials in Chapter 13; for the present you should learn *how* a diode works in a circuit, rather than *why* it works.

## 1.3   **Types of diode**

Figure 1.4 shows the structures of some types of diode. Figure 1.4(a) is a *point-contact* type. It has this name since a metal point presses onto a piece of germanium or silicon. The point-contact diode is used only in low-current circuits; for instance, the germanium point-contact diode type

OA91 is restricted to carrying a continuous current of 50 mA. It has the same outline as the OA202, but the OA202 is an example of a *silicon junction diode*. The word 'junction' refers to the fact that different types of germanium, or more usually silicon, are *diffused* into each other to produce a junction between them. The effect of this junction is explained in Chapter 16.

The ISJ50 is based on the semiconductor silicon, but it is made by a method known as the *epitaxial process*. The epitaxial structure is outlined later in this book where transistors are discussed. The 1N4000 series of diodes, figure 1.4(c), are designed to carry a current of 1 A and are useful in mains rectifier circuits. They are encased in moulded plastics and are of the diffused junction type, made of silicon and of *planar construction*.

The *maximum average forward* current which is allowed for a diode is published by the manufacturer of the diode. In the list of diode characteristics, you will see it as $I_F$ (av). For the germanium point-contact diode OA91, this is 50 mA. The *maximum surge forward current* which is allowed is indicated by $I_{fsm}$, and for the OA91 it is 150 mA. For the 1N4001, $I_F$ (av) is 1 A and $I_{fsm}$ is 30 A.

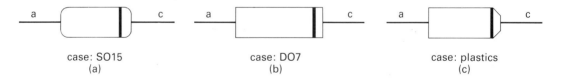

case: SO15
(a)

case: DO7
(b)

case: plastics
(c)

**Figure 1.4**  (a)  Examples: OA 81, OA 70   (b)  Examples: OA 90, OA 91, OA 200, OA 202   (c)  Examples: 1N 4001 to 1N 4007

# 2 Simple Experiments with a Diode

## 2.1 Experiment C1

### Testing a diode

Using an analogue multimeter, you can easily discover that a diode does not behave like an ordinary resistor. Switch a multimeter to its 'ohms × 1' range. Connect its leads across the diode as shown in figure 2.1.

## Question

1 What is the resistance of the diode? Does it seem to be high or low?

Reverse the leads to the diode and measure the resistance again. You will notice that the resistance is higher in one direction than in the other.

Suppose the first connection made was that shown in figure 2.1(a). Notice that the black lead is connected to the anode of the diode. Now you must remember that the *black lead* of an analogue multimeter has a *positive polarity*, because of the internal battery which is brought into operation when the multimeter acts as an ohmmeter. This positive polarity drives current through the diode in the direction of the arrow. This is the easy current direction for the diode, which is said to be *forward-biased*.

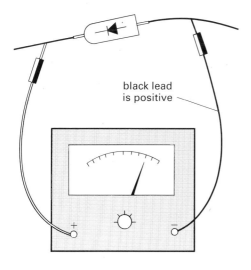

(a)

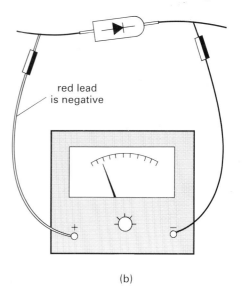

(b)

**Figure 2.1** Experiment C1: Testing a diode
    (a) Ohmmeter indicates low resistance
    (b) Ohmmeter indicates high resistance

Reversing the connections to the diode as shown in figure 2.1(b) puts the positive terminal of the battery in contact with the cathode of the diode. Very little current flows through the diode, as indicated by the high resistance, and the diode is said to be *reverse-biased*.

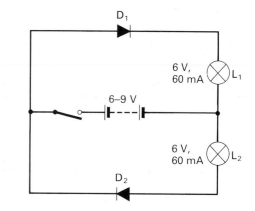

**Figure 2.3** Circuit 2

## 2.2 *Experiment* C2

# Controlling current flow with diodes

The following three circuits should be wired up on terminal block or breadboard. Use any general-purpose diodes such as OA91, 1N4148 or 1N4001.

### Circuit 1
The circuit of figure 2.2 shows two diodes connected 'back to back'.

**3**   Which lamp would light if the battery connections were reversed in this circuit?

### Circuit 3
**4**   Which lamps light before the switch is closed in the circuit shown in figure 2.4?

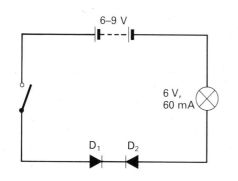

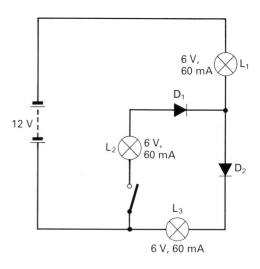

**Figure 2.2** Circuit 1

**Figure 2.4** Circuit 3

## *Questions*

**1**   Does the lamp light when the switch is closed?

### Circuit 2
**2**   Which lamp lights when the switch is closed in the circuit shown in figure 2.3?

**5**   Which lamps light after the switch is closed?

**6**   If the battery connections are reversed, which lamps light before and after the switch is closed?

# 3 The Voltage–Current Relationship for a Diode

## 3.1 Ohm's law and the diode

You know that a diode, unlike a resistor, is a one-way component. It should be interesting to discover whether Ohm's law applies to a diode as it does to a resistor. You must remember that, for Ohm's law to be true, the voltage across a resistor must be proportional to the current through the resistor. Thus, if a graph of current against voltage is plotted, a straight line is obtained.

Look at figure 3.1. This shows the graph obtained by plotting current against voltage for a wire-wound resistor. The straight line shows that Ohm's law is true over the range of voltage and current used. This line is known as a *voltage–current characteristic*.

## *Questions*

1 What is the value of the resistor used?
2 Under what conditions is Ohm's law true?

## 3.2 *Experiment* C3

### Obtaining a diode characteristic

A graph of $I$ against $V$ may be obtained for a diode in the same way as a similar graph is obtained for a resistor. Connections for a suitable circuit are shown in figure 3.2. Note that the diode is forward-biased. The meters may be two multimeters switched to appropriate ranges of amperes and volts.

## *Questions*

1 Should the voltmeter have a high or a low resistance?

Record the voltage across and the current through the diode for different values of $R$ as suggested in the table of figure 3.3. Plot a graph of $I$ against $V$. It is known as a *forward-biased characteristic* for the diode.

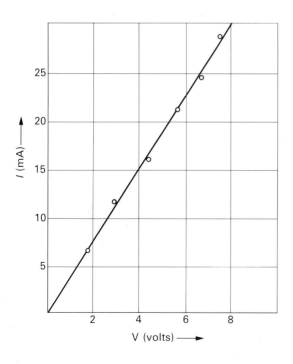

**Figure 3.1** Current–voltage characteristics for a wire-wound resistor

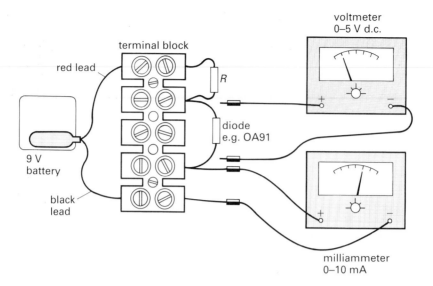

**Figure 3.2** Experiment C3: Obtaining a diode characteristic

| R (kΩ) | V (volts) | I (mA) |
|--------|-----------|--------|
| 100 | | |
| 82 | | |
| 47 | | |
| 33 | | |
| 27 | | |
| 22 | | |

**Figure 3.3** Table for recording results

2   Compare the graph in figure 3.4 with that of figure 3.1. Does the diode obey Ohm's law?

3   From your graph, determine the resistance of the diode for different values of $V$ and $I$. How does the resistance of the diode vary with the forward-bias voltage? These values are known as the *d.c.*, or *static*, *resistance* of the diode. The slope or gradient of the curve at a particular point is known as the *dynamic resistance* of the diode.

Note that the current is small for small forward voltages of less than about 0.1 V and then rises rapidly.

You know from measurements with an ohmmeter that if a diode is reverse-biased, a very small current flows because the ohmmeter reads a high resistance. Why not carry out an experiment similar to that above to obtain a reverse-biased characteristic? You could, but you would need a sensitive microammeter and a d.c. supply of about 25 V. What you would find is shown in figure 3.4. This graph is typical of what would be obtained for a silicon junction diode such as an OA202. You would find that the reverse current remains small and varies little as the reverse voltage increases. Eventually, however, the diode 'breaks down', which means that current flows through the diode in the reverse-bias direction. The *reverse breakdown voltage* $V_B$ varies from a few volts to a few hundred volts, depending upon the type of diode used. It is important not to exceed the reverse break-down voltage, since the diode is easily ruined by the large current which flows. However, use is made of the reverse breakdown voltage in the Zener diode, a special kind of diode described in Chapter 8. If you had a sensitive microammeter you could measure the reverse current that flows through the diode. The graph obtained for both the forward and reverse currents would be as shown in figure 3.4.

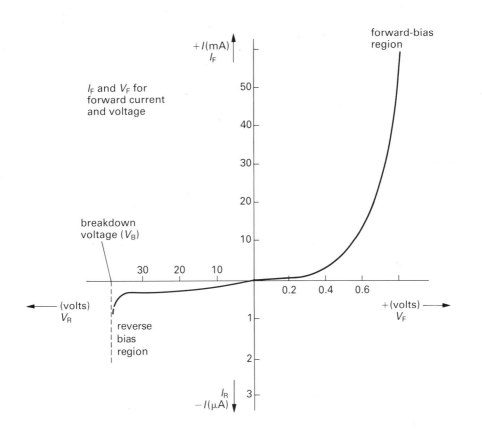

**Figure 3.4** Complete silicon-diode characteristic

# 4 Using Diodes for Rectification

## 4.1 Introduction

So far, you have experimented with diodes in d.c. circuits. Diodes are usually to be found in circuits through which a.c. flows, and a *rectification* circuit is one example.

You will need a low-voltage a.c. supply for the following experiments. Some workshops and laboratories have a transformer, the primary coil of which is already wound and may be plugged into the mains. The person needing a low-value a.c. voltage has only to wind a few secondary turns of wire over the encapsulation of the primary.

### CAUTION

In these experiments, and in the wiring up of the circuits to be described later, you must always take care not to be foolish when handling any equipment which is connected to the mains supply. Ensure that the mains leads of a device, such as a transformer, are completely insulated from you and from the box in which the transformer is mounted. If the box is of metal, then it should be connected to the earth lead from the mains power point. (See Book A, Section 8.5.)

## 4.2 Experiment C4

### Using an oscilloscope to show rectification

If you use an oscilloscope, it is possible to see what the diode is doing when it rectifies alternating current. Connections for a simple series circuit for this experiment are shown in figure 4.1. Use terminal block to assemble the circuit. Note the appearance of the trace on the scope. You will need to adjust the controls so that a stable trace is obtained — for help in this see Book A, Chapter 10.

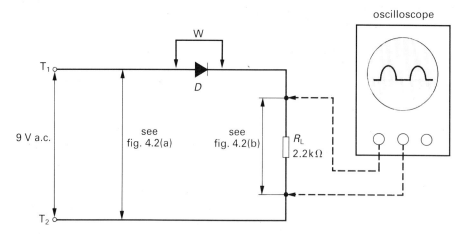

**Figure 4.1** Experiment C4: Using an oscilloscope to show rectification

## Questions

1   Reverse the diode. What changes occur in the trace?
2   Short out the diode using a wire link, W. What does the trace look like?

The oscilloscope measures the voltage across the load resistor $R_L$. This voltage is highest when most current flows through the diode and hence through the resistor. More current flows through the diode when it is forward-biased.

Now the a.c. supply is alternately forward-biasing and reverse-biasing the diode. It is forward-biased when the terminal $T_1$ is positive. It is reverse-biased when the terminal $T_2$ is positive. Since the supply is alternating at a frequency of 50 Hz, terminal $T_1$ will become positive 50 times a second. When the diode is shorted, the shape of the voltage changes of the a.c. supply can be seen on the oscilloscope. This shape is shown in figure 4.2(a). When the diode is in the circuit this shape is changed to that shown in figure 4.2(b). Note the change. Almost one half of the graph of the voltage change due to the a.c. supply has been removed. This occurs when terminal $T_2$ is positive and the diode is reverse-biased.

You will notice that, where the diode is reverse-biased, there is barely any trace on the oscilloscope as shown in figure 4.2(b). Here the resistance of the diode is very high so that no current flows through it.

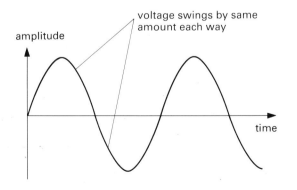

**Figure 4.2**  (a)  Unrectified waveform

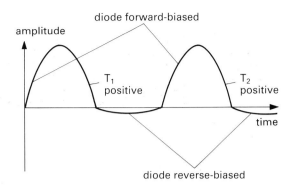

**Figure 4.2**  (b)  Rectified waveform

## 4.3 *Experiment* C₅

### Using diodes in a.c. circuits

**Experiment A**   Use the low-voltage alternating current supply to see what happens to the bulb in the circuit of figure 4.3. Terminal block may be used to assemble the circuit.

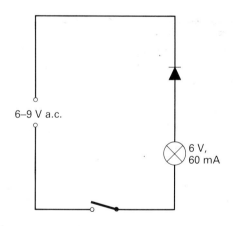

**Figure 4.3**  Circuit for Experiment A

This simple experiment shows how a diode rectifies an alternating current. It is true that this current is changing in amplitude, but it is a one-way or direct current. Before being able to use this direct current, however, the 'bumps' must be levelled out. This is known as *smoothing* the a.c., and is described in Section 5.3.

# *Questions*

**1** You should see that the bulb lights. Can you explain why it does? Is the bulb lit by direct or by alternating current flowing through it?

**Experiment B** Use terminal block and the a.c. supply to set up the circuit of figure 4.4.

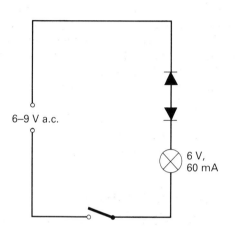

**Figure 4.4** Circuit for Experiment B

**2** Does the bulb light? Explain the result.

**Experiment C** Use terminal block and the a.c. supply to set up the circuit of figure 4.5.

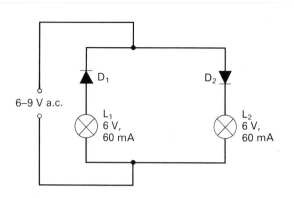

**Figure 4.5** Circuit for Experiment C

**3** Which bulb lights? Why?

**Experiment D** Look at figure 4.6. There are two questions about this circuit:

**4** Which bulb lights before you close the switch?
**5** Which bulb lights after you close the switch? Check your answer by wiring up the circuit and explain what you see.

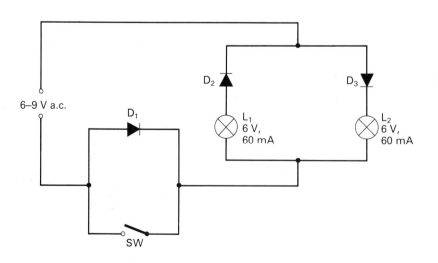

**Figure 4.6** Circuit for Experiment D

# 5 Half-wave Rectification

## 5.1 Introduction

As the previous experiments have shown, rectification is the conversion of alternating current to direct current.

Although the current through $R_L$ in figure 4.1 is a one-way current (d.c.), it is a very rough direct current. It reaches a peak value once every cycle of the alternating supply and is zero for half a cycle, that is, for 1/100 second. In order to smooth this current, a capacitor is used, as the next experiment shows.

## 5.2 *Experiment* C6

### Smoothing rectified a.c. supply

Assemble the circuit shown in figure 5.1 using terminal block so that different value capacitors can be connected across the load resistor, $R_L$, which should have a value of about 2.2 kΩ.

Beginning with a small capacitor of about 1 μF, connect this across $R_L$ and adjust the oscilloscope so that a good trace is obtained. Compare the traces obtained as you use capacitors of value up to 100 μF. You will be using electrolytic capacitors for some of these values, so take care to note the polarity of these and connect them so that their positive pole is connected to the cathode of the diode. Also, ensure that you are using capacitors whose working voltage is about 20 V.

As you increase the value of $C$, you will see that the average level of the trace will rise and become flatter.

## 5.3 How the capacitor smooths the rectified a.c.

The action of the capacitor is explained by means of figure 5.2. As the diode conducts when terminal $T_1$ becomes positive, $C$ charges up to the peak voltage of the a.c. supply. During this time, current is also passing through $R_L$.

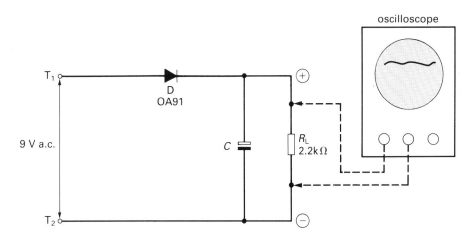

**Figure 5.1** Experiment C6: Smoothing a rectified a.c. supply

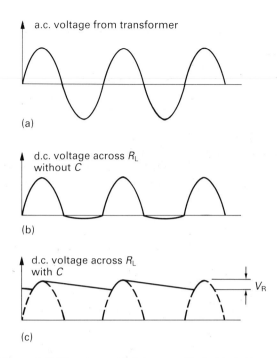

Figure 5.2 Waveforms for (a) unrectified, (b) half-wave rectified but unsmoothed and (c) half-wave rectified and smoothed sinusoidal voltages

When the voltage at $T_1$ begins to fall, current still continues to pass through $R_L$ since $C$ is discharging. If the time constant $(CR_L)$ is high enough, then $C$ takes some time to discharge. If the voltage on $C$ falls little during the time for the voltage at $T_1$ to fall to zero, go negative, and become positive again, then the d.c. is efficiently smoothed.

For a good smoothing action (that is, small ripple voltage $V_R$), $C \times R_L$ must be larger than 1/50 second. This is the time for the voltage on $T_1$ to go from one positive peak to the next positive peak. If $C = 100\ \mu F$, and $R_L = 1\ k\Omega$, $CR_L = 100 \times 10^{-6} \times 10^3 = 1/10$ second. This is five times greater than 1/50 second which means that the ripple voltage is small.

## Questions

1   Suppose $R_L$ falls to 200 $\Omega$. What is $CR_L$, and does this give poorer or better smoothing?

2   Are you clear why this is called 'half-wave' rectification?

## 5·4   Two practical points to note

There are two practical points to note about this half-wave rectifying circuit:

(a)   during the half cycle (1/100 second) when the voltage at $T_1$ changes from its maximum positive value to its maximum negative value, the voltage across the diode reaches almost twice the peak voltage of the supply. This happens because the capacitor discharges little during this time. Twice the peak voltage is called the *peak inverse voltage* ($V_{PN}$). For instance, the OA200 is rated to withstand a $V_{PN}$ of 50 V; the OA91, a $V_{PN}$ of 115 V; the 1N4002, a $V_{PN}$ of 100 V; and the 1N4004, a $V_{PN}$ of 400 V.

(b)   you must ensure that the diode can pass the current to the capacitor and to the load. This current should not exceed $I_F$ (av). You can calculate the current into the capacitor from the equation

$$I_{max} = V_{peak}/\text{capacitive reactance}$$

Since capacitative reactance = $1/2\ \pi fC$,

$$I_{max} = 2\pi fCV_{peak}$$
(see section Book B, Chapter 13).

If $V_{peak} = 12$ V, $C = 100\ \mu F$, and $f = 50$ Hz,

then

$$I_{max} = 3\ mA \text{ (approximately)}$$

To this must be added the current taken by $R_L$. For some diodes, the maximum power which is allowed to be dissipated is given. This may be calculated from the equation

$$P = I_F \text{ (av)} \times V_F$$

which is measured in milliwatts (mW) if $I_{F(av)}$ is measured in milliamps and $V_F$ in volts.

# 6 Full-wave Rectification

## 6.1 Introduction

The trouble with half-wave rectification is that current flows through $R_L$ (figure 5.1) during only half a cycle every complete cycle. This means that the full power of the a.c. supply is not being used. There are two commonly used *full-wave-rectifier* circuits which overcome this problem.

## 6.2 *Experiment* C7

### Designing a full-wave rectifier

Set up the circuit shown in figure 6.1. You may use point-contact diodes again, and $R_L$ should be about 2.2 kΩ.

Use an oscilloscope to examine the voltage variation across $R_L$. Compare it with the variation shown in figure 5.2(a). You will understand why this is called full-wave rectification. It is just as if the missing

halves of the half-wave circuit have appeared in the empty spaces, but inverted to give a voltage in the required direction.

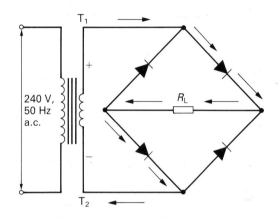

**Figure 6.2** Current flow for one half-cycle

Figures 6.2 and 6.3 will help you understand how this full-wave rectifier works. Terminals $T_1$ and $T_2$ are alternately positive and negative. When $T_1$ is positive (figure 6.2), current flows in the direction of the arrows. When $T_2$ is positive (figure 6.3),

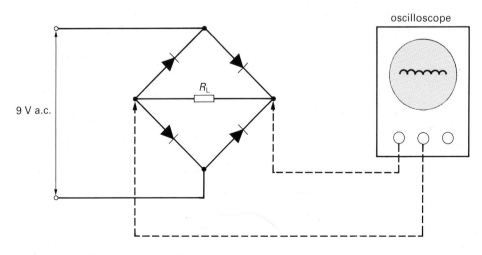

**Figure 6.1** Experiment C7: Designing a full-wave rectifier

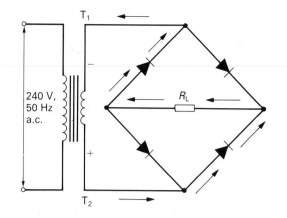

**Figure 6.3** Current flow for succeeding half-cycle

current flows in the direction shown. Notice that the direction of the current through $R_L$ remains the same regardless of the polarity of $T_1$. This is how the missing half-wave appears, as shown in figure 6.4.

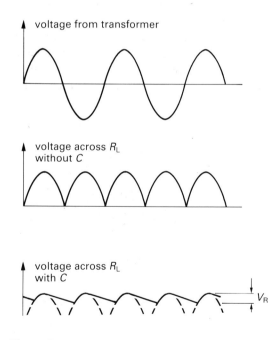

**Figure 6.4** Stages in full-wave rectification

Notice that the time between the peaks is half that for half-wave rectification. This helps the smoothing of the d.c., even before a capacitor is placed across $R_L$.

## *Questions*

1   Connect capacitors across $R_L$, and note their effect on smoothing as indicated by the ripple voltage $V_R$. Compare the traces you obtain on the oscilloscope with those shown in figure 6.4.

2   What is the ripple frequency produced by the full-wave rectifier circuit? Compare it with the ripple frequency produced by the half-wave circuit.

## 6.3   **Bridge rectifiers**

For full-wave rectification, it is not necessary to use four individual diodes since all four diodes can be obtained in one package as shown in figure 6.5. Each package has two a.c. input terminals and two d.c. output terminals marked + and −.

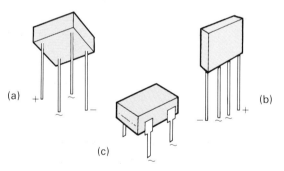

**Figure 6.5** Some examples of bridge rectifiers: (a) moulded  (b) in-line  (c) dual-in-line (DIL)

Bridge rectifiers are classified in the same way as single diodes, i.e. by their maximum average forward current, $I_{F (av)}$, and their maximum reverse voltage, $V_{RRM}$.

Chapter 9 describes how a bridge rectifier is used in a mains-operated power supply.

## *Question*

1   How would you use two diodes and a transformer which has a centre-tapped secondary winding. Draw the circuit.

# 7 A Simple Radio Receiver

## 7.1 What a crystal set was

Everyone at some time or other must have heard of the *crystal set*. This was an early type of radio which did not require batteries. There was a 'cat's whisker' (a fine piece of wire) which pressed onto a crystal (a small piece of lead sulphide, known as 'galena'). The would-be listener plugged in her earphones and adjusted the cat's whisker and the tuning capacitor to receive the required station. Eventually, radios using valves solved the problem of having to listen so hard, since they provided the necessary amplification — see Book A, Chapter 2. Transistors and integrated circuits now replace valves in modern radio receivers. However, a modern-day crystal set can be designed using a germanium rectifier diode as explained in Section 7.3, but first we need to know something about radio waves.

## 7.2 How radio signals are processed

Radio waves sent out from a transmitter travel at the speed of light (or of X-rays or of heat rays), a speed of $(3 \times 10^8)$ metres per second. These radio waves are *modulated* in one of two common ways: they are either *amplitude-modulated* (AM) or *frequency-modulated* (FM). Only AM waves will be explained here.

Figure 7.1 shows an amplitude-modulated carrier wave. The carrier wave has a high frequency; for instance, the frequency of the carrier wave transmitted from Radio Luxembourg is about 1.5 MHz (million hertz). This is obtained by dividing the speed of light by the wavelength of the station in metres, that is, $3 \times 10^8/208$.

Sometimes radio tuning dials are calibrated with frequency, not wavelength. Of course, the frequency of the modulation is in the audio range, which might go up to about 15 kHz, and is therefore much lower than the frequency of the carrier wave. You will find a fuller explanation of radio reception in Book D, Chapter 12.

Now it is not possible to hear the radio signal modulated as shown in figure 7.1. Earphones will not detect this kind of signal, since it has equal positive and negative swings and does nothing to the diaphragm in the earphone. The carrier wave has to be *detected* or *demodulated*, and this is where the diode is useful. If the signal shown in figure 7.1 is passed through a diode, one half of the wave is removed by the rectifying action of the diode, as shown in figure 7.2. Another part of the simple radio circuit then removes the carrier wave,

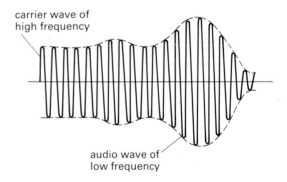

**Figure 7.1** The appearance of an amplitude-modulated carrier wave

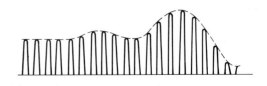

**Figure 7.2** A 'detected' radio wave

leaving only the audio-frequency wave of speech or music (figure 7.3) which operates the earphones.

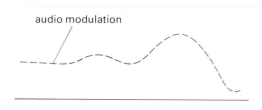

audio modulation

**Figure 7.3** Audio-frequency modulation

# 7.3 *Experiment* C8

## A design for a simple radio receiver

The radio receiver circuit shown in figure 7.4 uses a germanium diode, $D_1$, to detect amplitude-modulated radio frequency waves which are tuned in by the tuned circuit comprising coil, $L_1$, and variable capacitor, $VC_1$. The tuned circuit is described in Book B, Chapter 21. The npn transistor, $Tr_1$, amplifies the weak signals which are passed by the diode so that audio frequency signals are heard in the crystal earpiece, $EP_1$. The use of a transistor as an amplifier is described later in this book. Suitable transistors for this circuit are shown in figure 7.4. Capacitor $C_1$ is a coupling capacitor which isolates the tuned circuit from the d.c. voltage which just switches on the transistor. The amplified audio frequency signals appear across the resistor, $R_2$, in the collector circuit of $Tr_1$

which is why the earpiece is connected here.

The coil is made from 30 s.w.g. enamelled copper wire which is wound on a short length of ferrite rod. The coil can be wound first on a former of cardboard or plastic so that it can just be slipped over the rod. This coil is directional and should be turned so that the station you want is received at its loudest. Make sure that the enamel is scraped off the ends of the wire before connecting it into the circuit.

## Components required

Coil $L_1$: enamelled copper wire, turns as shown
Ferrite rod: about 30 mm long
Diode $D_1$: OA91 or similar germanium transistor
Capacitor $C_1$: 1 μF, 16 V electrolytic
Resistors $R_1$ and $R_2$: 100 kΩ and 1.2 kΩ, respectively 0.5 W, ±10% or less
Transistor: $Tr_1$ npn type
Earpiece: a crystal type or other high impedance earpiece
Switch $SW_1$: on-off type, slider or toggle action
Variable capacitor $VC_1$: 100 pF maximum value

It is not always necessary to use a high, long aerial to get good reception from this radio receiver. Try using the frame of a metal bookshelf. The shortcoming of this radio is that it is not very selective and a strong station will swamp weaker stations transmitting on wavelengths close to the strong station. See Project Module B5 for a much better radio receiver.

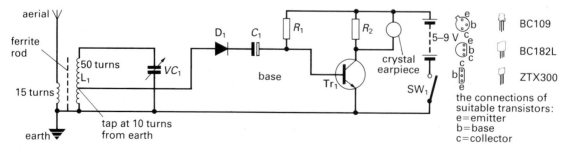

**Figure 7.4** A simple radio receiver

# 8 The Zener diode

## 8.1 What it does

This is a special kind of diode which is used to provide a *standard* or *reference* voltage. For this reason, the Zener diode is often referred to as a *reference diode*.

There are many circuits in electronics which only do their job properly if they are provided with a steady supply voltage; for example digital circuits (Book E). Due to its internal resistance, the terminal voltage of a battery changes as the current it delivers changes. A Zener diode can be used to stabilise this d.c. voltage so that a circuit is provided with a stabilised supply voltage. The Zener diode can be used as a discrete device in a circuit or it may be one of the components in a special integrated circuit called a *voltage regulator*. This chapter describes the use of Zener diode and Chapter 9 the voltage regulator.

## 8.2 What it looks like

From the outside, a Zener diode (usually called just a Zener) looks very much like an ordinary rectifier diode. Figure 8.1 shows the appearance of the very popular BZY88 series of 400 mW Zener diodes which are glass-encapsulated and have a DO35 outline. Others which dissipate more power have a different outline, e.g. the 1N5333 series can dissipate 5 W and have an outline similar to the 1N4000 series of rectifier diodes — see Section 1.3.

Zener diodes have a cathode and an anode terminal. The cathode is often marked with a band as for a rectifier diode. A Zener diode generally has its value marked on it. This value is the voltage it is able to stabilise; it might be marked '5V6' or '10 V' meaning that it can stabilise a 5.6 V supply or a 10 V supply as

explained below. The symbol for a Zener diode is

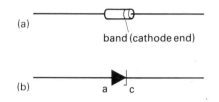

Figure 8.1 (a) Outline of the BZY88 Zener diode (b) Symbol of a Zener diode

## 8.3 The reverse-bias characteristic of a Zener diode

Figure 8.2 shows what the voltage–current characteristic of a Zener diode looks like. Its shape is similar to that of a rectifier diode. The Zener diode is not allowed to operate in the forward-bias direction where the current increases sharply as for a rectifier diode. Instead it is designed to operate in the reverse-bias breakdown region where a rectifier diode would normally suffer permanent damage.

Figure 8.2 (overleaf) shows that a large current flows from cathode to anode when the reverse-bias voltage reaches the breakdown voltage of 9 V. The graph shows that the current changes dramatically in the breakdown region without the voltage across the diode changing very much. This is the secret of the Zener diode, for a large change of current through the diode requires only a small change of voltage across the diode, provided the Zener diode is operated in the breakdown region.

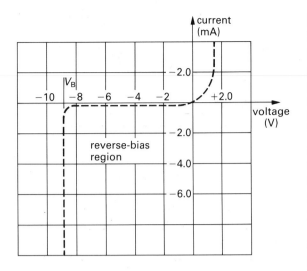

**Figure 8.2** The important reverse-bias region of a Zener diode characteristic

## 8.4 *Experiment* C9

# Finding out what a Zener diode does

Suppose you want to use a Zener diode to provide a steady voltage of about 7 V across a load, even though the load resistance changes. The circuit of figure 8.3 is one simple way of providing this voltage from a d.c. supply, the voltage of which may change.

Firstly, you must choose a Zener diode which gives the required reference voltage.

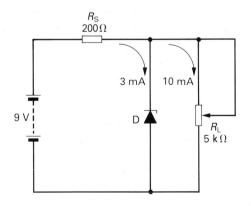

**Figure 8.3** Experiment C8: Finding out what a Zener diode does

For this experiment, a BZY88 400 mW, 6V8 Zener diode would be suitable. This diode has a breakdown voltage of 6.8 V, and it should not be allowed to dissipate more than 400 mW. Connect together the components of the circuit shown in figure 8.3, using breadboard or terminal block. You will need two multimeters, one switched to its 10 V (or nearest) range and the other switched to its 5 mA (or nearest) range. $R_L$ is a 5 k$\Omega$ variable resistor and represents a load, the resistance of which you can vary easily.

Adjust $R_L$ and watch the meters carefully. You will notice that the voltage across $R_L$ (which is also the voltage across D) remains steady at 6.8 V, even though the load resistance changes; that is, even though the current drawn by the load changes. Adjust $R_L$ to the point where you just see the voltage dropping below 6.8 V. Remove $R_L$ and measure its resistance, for this setting, using a multimeter. This is the minimum value of the load resistance for which the diode provides a steady voltage of 6.8 V.

Thus the Zener diode provides a stable reference voltage, even though there are changes (within limits) of the supply voltage and of the load resistance.

▽ 8.5 **Power dissipation in a Zener diode**

The purpose of the series resistor $R_S$, is to 'drop' the surplus voltage not required by the Zener diode. In the case of the circuit in figure 8.3, $R_S$ drops $(9 - 6.8)$ V = 2.2 V. This resistor must be able to dissipate the power developed in it. The power is easy to calculate if you know the current through the Zener diode and through the load. For instance, suppose the current through $R_L$ is 10 mA and through the diode is 3 mA, as shown in figure 8.3. You will see that the current through $R_S$ is 13 mA.

From the equation $P = V \times I$,

$$P = 13 \text{ mA} \times 2.2 \text{ V} = 28.6 \text{ mW}.$$

This is well within the power limit of a $\frac{1}{4}$ W △ resistor.

# 9 Practical Mains-operated Power Supplies

## 9.1 Introduction

Electronic equipment which is to operate remotely, for instance in a spacecraft or on a weather ship, has to be powered by some sort of battery. But if the equipment, e.g. a radio broadcast transmitter, requires a lot of power, a mains power supply is generally provided. A mains power supply is usually more convenient than batteries in the home or workshop since it guarantees a steady supply voltage without having to replace batteries. A suitable mains-operated power supply is shown in figure 9.1.

As figure 9.2 shows, there are four main building blocks in the power supplies described in this chapter. Building block 1

**Figure 9.1** A mains-operated power supply

is a transformer which converts the 240 V a.c. supply into a lower, e.g. 15 V, a.c. supply. Building block 2 uses rectifier diodes to change the low-voltage a.c. into a low voltage half-wave or full-wave rectified d.c. Building block 3 uses a capacitor to smooth the rectified a.c. Building block 4 is optional and uses a Zener diode or a voltage regulator to stabilise the rectified d.c. voltage so that the d.c. voltage remains steady for a wide variation in mains supply voltage and load current.

### CAUTION

If you are unsure about connecting any equipment to the mains supply, do ask for advice from someone who has had experience of handling or servicing mains-operated equipment. The mains supply can produce a lethal electrical shock.

## 9.2 A simple d.c. power supply

Figure 9.3 shows a d.c. supply which provides about 9.8 V at 1.5 A (max) suitable for replacing the battery in most types of transistor radio. This supply can be

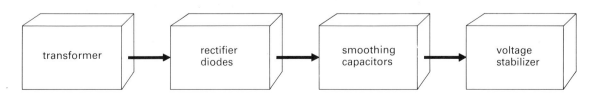

**Figure 9.2** Block diagram of a mains-operated power supply

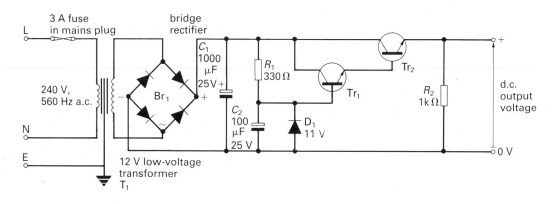

**Figure 9.3** Circuit diagram of the simple power supply

used to power most of the Project Modules described in each book of *Basic Electronics*.

Components which have the following specifications should be used in this power supply.

Transformer, $T_1$:   20VA type, two 12 V, 0.8 A secondary windings

Bridge rectifier, $Br_1$:   1.6 A, in-line type or similar

Capacitors, $C_1$ and $C_2$:   electrolytic types, 1000 μF, and 100 μF, respectively, both with a working voltage of at least 25 V

Resistors, $R_1$ and $R_2$.   330 Ω and 1 kΩ, respectively, 0.5 W, ± 10% or less

Zener diode:   11 V, 400 mW maximum power dissipation

Fuse:   2 A rating

Transistors $Tr_1$ and $Tr_2$:   2N3053 and TIP31, respectively

Materials:   terminal block for assembling the circuits; fuse holder; PVC-covered stranded wire for connections; 3–core mains cable and plug fitted with 3 A fuse; box for housing circuit; on-off mains switch and sockets for d.c. power output.

The transformer has two identical secondary windings, each rated at 12 V, 0.8 A. These windings should be connected in parallel so that the maximum current which the transformer can deliver is about 1.5 A. You should make sure that the current which you draw from the power

supply does not exceed this current maximum, otherwise the bridge rectifier or transformer may be damaged.

Most of the components for this power supply can be assembled on a short length of terminal strip as shown in figure 9.4. Connections to the transformer should be made using the PVC-covered stranded wire. An on-off switch is optional but, if used, should isolate the mains supply to the transformer.

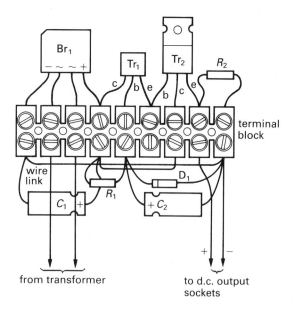

**Figure 9.4** Assembly of the power supply on terminal block

The transformer should be earthed to the earth lead of the mains cable using a solder tag bolted to one of the feet of the transformer. Stranded wire should be used to connect the output of the power supply to two 4 mm sockets mounted on the box which houses the power supply. If a metal box is used, it should be earthed to the mains supply as explained in Book A, Section 8.5. Use nylon screws to fix the terminal block firmly to the base or side of the box. Make sure that the components are neatly arranged as shown in figure 9.4. Use PVC-covered wire for the wire link and ensure that component leads cannot short together or touch the box if it is made of metal.

## 9.3 Using a fixed-voltage regulator

The construction of the simple d.c. power supply described in Section 9.2 can be simplified by using a three-terminal voltage regulator to replace the Zener diode and the two transistors. The regulator has the advantage that if too much current is drawn from the power supply, it will automatically shut down. In this way no damage can be done to the regulator or to the transformer by overheating.

Voltage regulators accept an unstabilised d.c. voltage and produce a stabilised voltage usually of 5 V, 12 V or 15 V. Figure 9.5 shows the appearance of three types of voltage regulators. Note that each regulator has three terminals labelled input (i/p), common (com) and output (o/p). Some circuits such as those based on operational amplifiers (see Book D, Chapter 5) need to be operated from both a positive and negative power supply. The voltage regulators known as the '79 series' and the '79L series' are designed to stabilise a negative supply.

Figure 9.6 shows how these fixed voltage regulators are used. The low voltage a.c. from the transformer is rectified by the bridge rectifier, $Br_1$, and smoothed by the capacitor, $C_1$, as for the simple power supply (figure 9.3). The voltage regulator replaces the Zener diode and transistors in the simple power supply. Capacitors $C_1$ and $C_2$ make sure that the voltage regulator is stable in operation.

Like the simple power supply, this regulated voltage power supply can be assembled on a length of terminal block. The following transformers and bridge rectifiers are recommended for use with the 1 A plastics regulators:

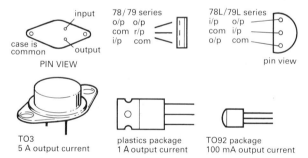

**Figure 9.5** Three types of voltage regulators

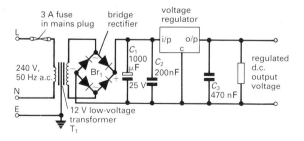

**Figure 9.6** The use of a voltage regulator in a power supply

| regulator | transformer |
| --- | --- |
| 7805 (5 V) | two 4.5 V, 2.2 A windings connected in series |
| 7812 (12 V) | two 15 V, 1.6 A windings connected in parallel |

The regulators should be bolted to a heatsink as recommended on the manufacturer's data sheet. The bridge rectifiers should be rated at 1.6 A.

# $10$ The Voltage-doubler Circuit

## 10.1 A diode and capacitor in series

Figure 10.1 shows a circuit in which a capacitor is being charged through a diode. As you know, the diode conducts easily when terminal $T_1$ is positive. This means that $C$ charges to almost the peak voltage of the supply.

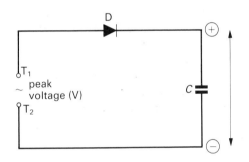

**Figure 10.1** Basic diode – capacitor charging circuit

## 10.2 How a voltage-doubler works

Look at the figure 10.2. This circuit consists of two circuits like figure 10.1 connected in parallel. What is the size of the output voltage? This question can be answered by regarding the current as flowing in two loops.

During one half-cycle of the supply voltage. $C_1$ charges. During the next half-cycle, $C_2$ charges. Each capicator charges to a little less than the peak voltage of the supply. These two voltages are in the same direction and, added together, give about △ twice the input voltage.

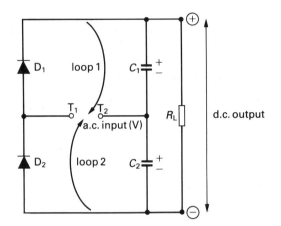

**Figure 10.2** A voltage-doubler circuit

## 10.3 Another voltage-doubler

Figure 10.2 shows an alternative voltage-doubler circuit. It too can be considered to consist of two loops. When $T_2$ is positive, $C_1$ charges, since current flows through loop 1. During the next half-cycle, $C_2$ charges to the combined voltage of the supply and that across $C_1$. Thus the voltage across $R_L$ approaches twice the voltage across the input. In practice, the voltage is less than twice the peak input voltage, and △ it varies as the load varies.

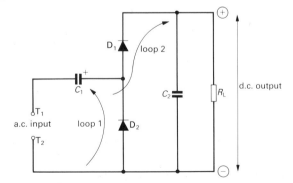

**Figure 10.3** Another voltage-doubler circuit

# 11 The Light Emitting Diode and Photodiode

## 11.1 Introduction

The rectifier diode and the Zener diode are made by creating a junction (a p-n junction) between two types of semiconductor (n-type and p-type). As explained in Chapter 6, this p-n junction acts as a one-way valve for current flowing through it.

By using specially selected p-type and n-type semiconductors, it is possible to make the p-n junction emit coloured light to produce a *light emitting diode* (LED). The light is emitted when current flows through the LED in the forward-bias direction. The LED is used as a low current indicator lamp in many types of consumer and industrial equipment, such as hi-fi systems and machinery control panels. As explained in Book E, the LED is also used in seven-segment displays to show numbers and letters. By selecting the right type of 'impurity' in the semiconductor compound (usually gallium arsenide phosphide), the colour of the light emitted can be made red, yellow, green or blue. The LED does not emit light when it is reverse-biased.

In the *photodiode* a small current flowing in the reverse-bias direction varies with the amount of light which reaches the junction. The photodiode is used in camera light-meters to measure light intensity, and also for rapid counting of the rotation of wheels in control equipment. The photodiode is able to respond much faster to light changes than the light dependent resistor (LDR).

## 11.2 The light emitting diode

As figure 11.1 shows, the LED is designed to emit light when current flows through it

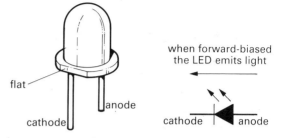

**Figure 11.1** General appearance of and symbol for a light emitting diode

from its anode to its cathode terminal. In a common version of the LED, the cathode terminal is indentified by the *flat* near the base of the LED. The Project Modules in *Basic Electronics* use this type of LED as an on/off indicator in counters and switches. Infrared LEDs are extensively used in optical communications systems — see Project Module B6.

An LED only needs about 2 V across its anode and cathode terminals to make it emit light. If a higher voltage is used, the current which flows through it may be high enough to damage it.

In order to limit the current when an LED is used at higher voltages, a resistor must be connected in series with it. Figure 11.2 shows how to calculate the value of

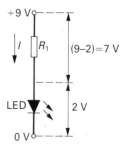

**Figure 11.2** Calculating the value of the series resistor for an LED

this series resistor, $R_s$. Suppose you want to operate the LED from a 9 V supply voltage, $V_s$. If the current, $I$, which flows through the LED is to be limited to 10 mA, the value of the series resistor is calculated as follows:

voltage across $R_1 = (V_s - V_f) = (9 - 2)$
$= 7$ V
current through $R_1 = 10$ mA $= 7V/R_1$
therefore $R_1 = (7V/10$ mA$)$
$\qquad = 700$ ohms

A series resistor of value 680 $\Omega$ would be suitable. However, an LED emits light safely for any current in the range 5 mA to 25 mA. Thus the series resistor could have any value between about 1.5 k$\Omega$ and 270 $\Omega$. If the current is allowed to increase above 25 mA, the life of the LED will be reduced.

Figure 11.3 shows two ways of using an LED to indicate the output state of a digital logic circuit. In the first case, if the LED is on, the digital output is high; if it is off, the digital output if low. In this example, the logic circuit is said to *source* the current required by the LED. In the second case, the logic circuit *sinks* the current required by the LED. If the LED is on, the digital

output is low; if it is off, the digital output is high.

If the logic circuit is unable to source or sink the current required by the LED, a transistor can be used as shown in figure 11.3. A small current to the base of the transistor switches on a larger current which flows through the LED direct from the power supply.

## 11.3 The photodiode

The appearance of a typical photodiode is shown in figure 11.4. It has two terminals, anode and cathode and a flat window at its end through which light reaches its p-n junction. A photodiode is operated in reverse-bias as shown. The small reverse current which flows through it varies in proportion to the amount of light reaching the junction. A photodiode is used in Project Module B6 as a detector of infrared light in an optical communications and control system, respectively. This application relies on the fast response of the photodiode to a change of light intensity.

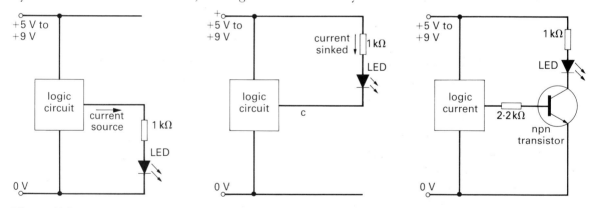

**Figure 11.3** Three ways of using an LED with switching circuits

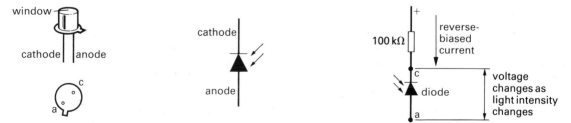

**Figure 11.4** The appearance, symbol and use of a typical photodiode

# 12 Resistivity and Conductivity

## ▽ 12.1 Conductors and insulators

Using a multimeter switched to its highest resistance range you can easily find out which materials conduct and which resist the flow of electricity.

For example, if you test metals, you will find that their resistance is very low — they are good electrical conductors. But test Perspex, polythene, glass, and nylon and you will find that their resistance is very high — the ohmmeter will show an infinitely high resistance.

## *Question*

1  In which group, conductors or insulators, would you place each of the following materials: copper, carbon, wood, leather, silver, air, water, △ aluminium?

## ▽ 12.2 The meaning of resistivity and conductivity

However, the resistance of a piece of material depends not only on what it is made of, but on such things as its temperature, length, and cross-sectional area. In order to compare the electrical-conducting properties of materials, they must be under identical conditions and the same shape and size. This is usually done by comparing the resistances of cubes of material of unit size, say 1 metre cubes, as shown in figure 12.1. In fact, the resistance between opposite end faces of such a cube is known as the *resistivity* of the material.

Of course, the resistivity of a material is not measured by obtaining a 1 metre cube

of it. Instead the resistance, the length and the area of cross-section of the material in the form of a wire is measured. The resistivity is then given by the equation

$$\text{resistivity} = \frac{\text{resistance} \times \genfrac{}{}{0pt}{}{\text{area of}}{\text{cross-section}}}{\text{length}} \; (\Omega)$$

Resitivity has units of ohm × metre ($\Omega$ m) since

$$\frac{\text{ohm} \times \text{metre}^2}{\text{metre}} = \text{ohm} \times \text{metre}$$

As well as resistivity, the term *conductivity* is used to describe the electrical-conducting properties of materials. Conductivity is the inverse of resistivity so that:

$$\text{conductivity} = \frac{1}{\text{resistivity}}$$

Therefore a material of high resistivity has a low conductivity. Thus the units of conductivity are

$$\frac{1}{\text{ohm} \times \text{metre}} = \Omega^{-1} \, \text{m}^{-1}$$

But $\Omega^{-1}$, the reciprocal ohm, is named the *siemen*. Thus the unit of conductivity is siemen/metre or $\text{s m}^{-1}$.

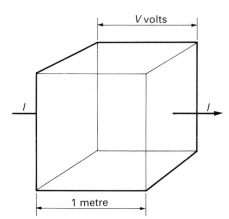

**Figure 12.1** An imaginary 1 metre cube for obtaining resistivity — Resistivity = $V/I$

## Questions

**1** Name two materials having high conductivity and two having high resistivity.
**2** Is the conductivity of a tungsten-filament lamp lower when it is off than when it is on?

Figure 12.2 compares the resistivities of different materials. Note that a good conductor like copper has a resistivity of about $10^{-7}$ $\Omega$ m. But polythene has a resistivity of about $10^{12}$ $\Omega$m, in the region △ of a million billion times higher!

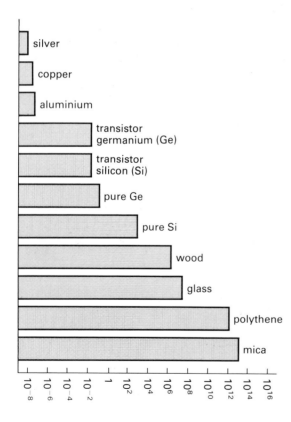

**Figure 12.2** A comparison of the resistivities of some materials

## ▽ 12.3 **Semiconductors**

These are 'mid-way' conductors, since they have a conductivity higher than insulators but lower than metals. Semiconductors are very important in the field of electronics, since they are used to make a whole range of modern semiconductor devices including the transistor, the diode and the thermistor.

Two of the commonest semiconductors are the elements silicon and germanium. At low temperatures, these happen to be good insulators; at everyday temperatures, say about 20°C, they will conduct some current and for higher temperatures, their conductivity gradually increases. Figure 12.2 shows that the resistivities of pure silicon and germanium are about mid-way between the insulators and conductors.

The way in which heat is responsible for increasing the conductivity of semiconductors is explained in Chapter 14. It also explains how a marked increase in the conductivity of silicon and germanium is brought about by the addition of carefully controlled amounts of impurity, specially selected to give p-type and n-type materials which are required for semiconductor devices. Figure 12.2 shows the slightly lower resistivities (higher conductivities) of such 'transistor' germanium and silicon.

Not all semiconductors are based on the elements silicon and germanium. It has been necessary to find some new ones. In Chapter 7, Book B, you used a light dependent resistor for which a semiconducting material known as *cadmium sulphide* is used. It has a conductivity which falls markedly when exposed to light. In Chapter 7 of Book B you also used a *thermistor*, which is a mixture of semiconducting oxides. These semiconductors generally consist of a chemical compound of two elements, such as a metal oxide or metal sulphide. And, not least, a substance called gallium arsenide is the basis of new semiconductors △ — see Book E, Chapter 15.

# 13 Materials for Making Semiconductors

▽ 13.1 ## The atomic structure of silicon and germanium

These are two very important materials from which semiconductors are made. It will help you to understand how semiconductors do their job if we have a look at the model of a silicon atom shown in figure 13.1. If you can remember the simple ideas about atomic structure you learnt in Book A, Chapter 7, you will be able to answer the following questions.

## Questions

1  How many protons are there in the nucleus of the silicon atom?
2  What kind of electric charge is carried by the electrons and by the protons? What can you say about these two charges?
3  Look at a periodic table of the elements and identify the position of germanium (Ge) and silicon (Si). What other elements lie in the same group as these two? Do you know what these elements have in common?

You will see from figure 13.1 that a silicon atom has four electrons in its outer shell. The same is true of germanium. Now, in a piece of silicon, all the atoms are held together by *bonds* which exist between the electrons. Each electron forms what is known as a *covalent bond* with one electron from a neighbouring atom, as shown in figure 13.2. These covalent bonds use up every one of the outer or *valence* electrons. In fact, in a piece of pure germanium or silicon, all the valence electrons are paired-

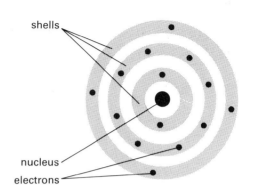

**Figure 13.1** An atomic model for a silicon atom

off by being used in these bonds.

As you will remember from Chapter 7, Book A, only those substances which have some weakly-bound electrons are electrical conductors. In metals, there are many electrons free to move from atom to atom. They are known as *conduction electrons*, since they move easily through the metal when a voltage is applied across it.

## Question

4  Would you expect a piece of pure germanium or silicon to be an electrical conductor or an electrical insulator?
△

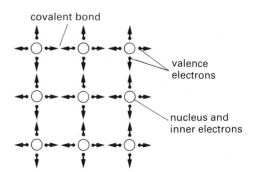

**Figure 13.2** The regular atomic structure of a silicon atom

# 14 Modifying the Conductivity of Germanium and Silicon

## ▽ 14.1 Silicon and germanium are insulators

Since there are no free electrons in a piece of pure silicon or germanium, you would expect these elements to be insulators. They are. They have a high resistivity, but there are two ways by which their resistivity can be decreased:

(a) by heating, and

△ (b) by adding impurities.

## ▽ 14.2 The effect of heating a semiconductor

When germanium or silicon is heated, the atoms vibrate more energetically, and this results in the breaking of some covalent bonds and electrons being released for conduction. A higher temperature breaks more bonds, more electrons are freed, and the conductivity increases further.

## *Question*

**1** Do you think that the conductivity of germanium or silicon increases or decreases as the temperature rises?

The name given to the increase of conductivity of silicon or germanium due to heat is *intrinsic conductivity*. Intrinsic means 'a natural part of'. In other words intrinsic conductivity is conductivity caused by electrons produced from the atoms of germanium or silicon — not from
△ atoms added from outside.

## ▽ 14.3 Adding impurities to silicon and germanium

When other atoms are introduced into a piece of germanium or silicon, the germanium and silicon become impure. The whole secret of the behaviour and usefulness of a semiconductor is the controlled introduction of impurity atoms into germanium and silicon so as to alter the conductivity in a known way. This process is known as *doping*.

Impurity atoms generally make a 'bad fit' into the regular atomic structure of pure germanium or silicon, due to the impurity atom having too many or too few valence electrons. Depending on the impurity, two kinds of semiconductor are
△ produced in this way — p-type or n-type.

## ▽ 14.4 The n-type semiconductor

An n-type semiconductor is produced by doping silicon or germanium with, for example, atoms of phosphorus. A phosphorus atom has five electrons in its outer shell and is therefore called a *pentavalent* atom. Figure 14.1 shows what happens when an atom of phosphorus becomes embedded in the atomic structure of silicon. Four of its valence electrons each share an electron with a neighbouring silicon atom, but the fifth cannot make a covalent bond. What happens to this electron? Since it is relatively weakly attached to its parent atom, it can wander about. Phosphorus is said to be a *donor impurity*, since it can donate (give away) an electron for conduction.

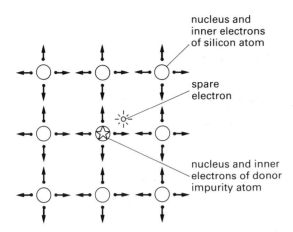

**Figure 14.1** How a donor-impurity atom provides an electron

valence electrons each share an electron with a neighbouring atom of silicon. There is thus one unsatisfied electron for every impurity atom. This electron is not available for conduction, but it will accept another electron to pair with it. The vacancy which remains to be filled is called a *hole*. Since this hole attracts an electron, it behaves as if it had a positive charge. Boron is said to be an *acceptor impurity*, since its atoms can accept an electron from other atoms nearby.

## Questions

**1** Why do you think silicon containing boron is called a 'p-type' semiconductor?
**2** Despite the presence of positively charged holes, why is p-type material electrically neutral?

## Questions

**1** Why do you think silicon containing phosphorus atoms is called an 'n-type' semiconductor?
**2** Despite the presence of electrons, why is n-type material electrically neutral?

## ▽ 14.5 The p-type semiconductor

A p-type semiconductor is produced by doping silicon (or germanium) with, for example, atoms of *boron*. A boron atom has three valence electrons in its outer shell and is therefore called a *trivalent* atom. Figure 14.2 shows what happens when an atom of boron becomes embedded in the atomic structure of silicon: three of its

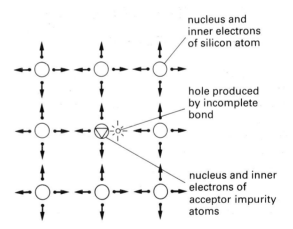

**Figure 14.2** How an acceptor-impurity atom provides a hole

# 15 Current Flow through a Semiconductor

## ▽ 15.1 Electron flow in an n-type semiconductor

Look at figure 15.1. It shows a cell connected across a piece of n-type semiconducting material. From what has been said in Section 14.5 you will understand that it is the weakly-bound spare electrons from the donor-impurity atoms which move as conduction electrons through the semiconductor. Electrons enter the material from the battery, but they do not move through the material smoothly, since impurity atoms are continually gaining and losing an electron. However, the general drift of these electrons is in the direction shown.

## *Question*

1 Do you expect as many electrons to leave the material as enter it in one △ minute, say?

## ▽ 15.2 Hole flow in a p-type semiconductor

Look at figure 15.2. It shows a cell connected across a piece of p-type semiconducting material. From what has been said in Section 14.6, you will understand that holes are created in the material because of the presence of

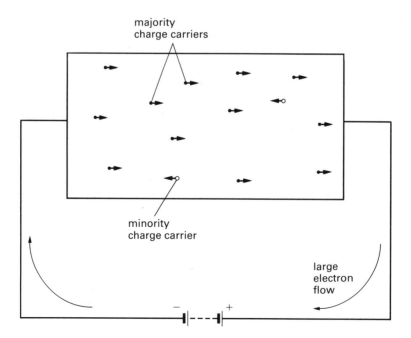

**Figure 15.1** Electron flow through an n-type semiconductor

acceptor-impurity atoms. Electrons move through the material alternately filling and leaving these holes. Current flow in p-type semiconductors is most conveniently considered to be due to the movement of positively charged holes. The flow of holes can be likened to the movement of an empty seat in a row of cinema seats.

Imagine an empty seat at the end of a row of seats in a cinema. The person next to the vacant seat moves into it, leaving a vacant seat into which, in turn, the next person moves, and so on. Which way does the empty seat move? In the case of semiconductors, the people correspond to the electrons and the empty seat to the holes.

## Question

**1** In figure 15.2, do as many electrons leave the p-type semiconducting material as enter it?

▽ 15.3 **Impurities have a low concentration**

It is important to note that the impurity atoms which are added to germanium and silicon are a very small fraction of the number of germanium and silicon atoms. The minute trace of one impurity atom to one million silicon atoms produces the increased conductivity which makes silicon so important to the making of present-day diodes, transistors, and other semiconductors.

The conduction electrons in n-type material and the conduction holes in p-type material are called *majority charge carriers*. In both p- and n-type material, a few electrons and holes are present from the broken covalent bonds between silicon or germanium atoms. These are called *minority charge carriers*.

## Question

**1** What are the minority charge carriers in p-type material and in n-type material?

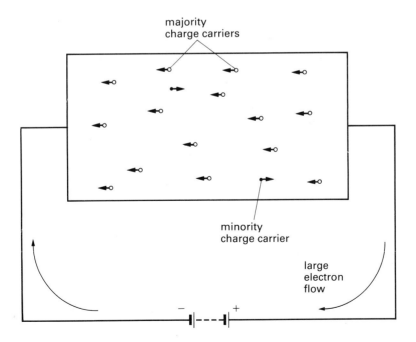

**Figure 15.2** Hole flow through a p-type semiconductor

# 16 The p-n Junction

## ▽ 16.1 An unbiased p-n junction

A p-n junction is generally produced when pieces of p-type and n-type material are joined together as shown in figure 16.1. Immediately after this junction is formed, there is flow of electrons from the n-type material across the junction to the p-type material, and a reverse flow of holes, as shown in figure 16.1.

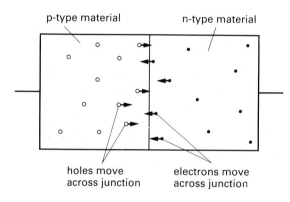

**Figure 16.1** Initial hole and electron movement across a p-n junction

The reason for the holes and electrons flowing in the directions they do should be clear from what has previously been said about the properties of p-and n-type materials. Electrons move to fill holes, and holes move to capture electrons. But this movement does not continue indefinitely. A negative charge builds up in the p-type material and a positive charge in the n-type material, and these prevent any further movement of charge. These charges produce a voltage across the junction in the direction of the arrow shown in figure 16.2. This is called the *junction voltage* which is about 0.1 V for germanium and 0.6 V for silicon.

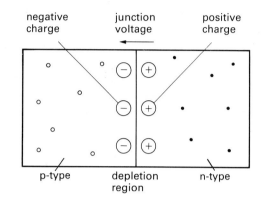

**Figure 16.2** Steady hole and electron distribution across an unbiased p-n junction

Because the flows of electrons and holes across the junction neutralise each other, there is produced a very narrow region known as the *depletion region*. This means that this region is depleted of electrons and holes. In fact, this depletion region has almost the high resistivity which you might expect from a piece of pure silicon or germanium.

Of course, the situation shown in figure 16.2 is altered when the diode is used. As you have seen, the diode is used in circuits either forward-biased (easy current flow) or reverse-biased (difficult current flow). We shall now have a look at the behaviour △ of the p-n junction under these conditions.

## ▽ 16.2 A forward-biased p-n junction

Figure 16.3 shows a battery connected so that the junction diode is forward-biased.

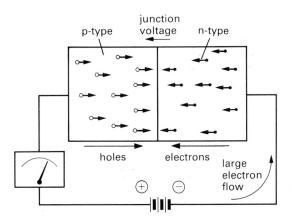

**Figure 16.3** Electron and hole movement across a forward-biased p-n junction

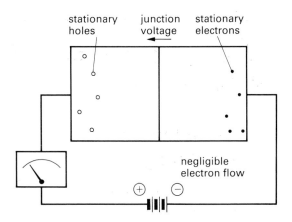

**Figure 16.4** Electron and hole distribution across a reverse biased p-n junction

## Questions

1 Which material is connected to the positive terminal of the battery?
2 Be quite sure you understand why electrons and holes flow in the directions shown. Note that hole flow is in the direction of the arrow on the diode symbol. Which is the conventional current flow direction?

The main point to note is that, before current can flow across the p-n junction in the easy current direction, the external voltage must first overcome the small opposing junction voltage generated across △ the junction as explained above.

▽ 16.3 **A reverse-biased p-n junction**

Figure 16.4 shows a battery connected so that the junction diode is reverse-biased.

## Question

1 Does the external battery oppose or aid the small internal junction voltage across the junction?

You should see that applying a reverse bias repels both electrons and holes away from the junction. The depletion region widens, and no current flows in the external circuit.

The final point to note is this: the above four diagrams are concerned with majority charge carriers only; that is, electrons in n-type material and holes in p-type material. But, as you know from Section 15.3, minority carriers exist, due to thermally generated electrons and holes in both types of material. These carriers can cross the junction in figure 16.4 and produce a small reverse current — a current which is lower in silicon diodes than in germanium diodes, and which increases with temperature and reverse voltage. The characteristics of the diodes listed in the catalogues of suppliers of electronic components include the reverse current for specified voltages.

Do remember, that although the reverse-bias current is of no use in a rectifier diode, it is the basis of operation of the Zener diode and the photodiode — see △ Chapters 8 and 11, respectively.

# *17* Transistors — An Introduction

## 17.1 Two sorts of transistor

Despite the widespread use of integrated circuits, the discrete transistor is still in great demand by the circuit designer. The use of the transistor as an *amplifier* and as a *switch* is explained in this book.

Since the transistor is capable of amplifying a signal, it is said to be an *active* components. Devices such as resistors, capacitors, inductors and diodes are not able to amplify and are therefore known as *passive* components.

There are two types of transistor in use:

**(a)** the *bipolar* or *junction transistor*

**(b)** the *unipolar* or *field-effect transistor*

Usually when people talk about the 'transistor', they mean a discrete bipolar transistor since this type is more widely used than the unipolar type. Bipolar means that the transistor's operation depends on the flow of both electrons and holes — see Chapter 23. Unipolar means that the transistor's operation depends on the flow of electrons or holes but not both — see Chapter 27.

## 17.2 What bipolar transistors look like

There are thousands of different types of bipolar transistor. Figure 17.1 shows what a few commonly-used types look like.

The case which protects the actual transistor is known as the *encapsulation*. The two encapsulations in widespread use are *metal* and *plastics*. The shape, or outline, of the transistor is chosen by the manufacture and depends on what it is to be used for.

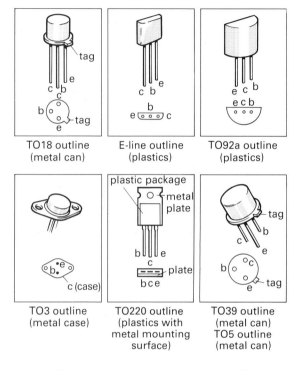

**Figure 17.1** The outlines of some common silicon transistors (underside views shown)

Each transistor is given a code which identifies it. American manufacturers use a code which begins with '2N' (for diodes the code begins '1N'). European manufacturers use a code which indicates the type of semiconductor, silicon or germanium, which the transistor is based on, and the intended use of the transistor. For example, for the BC108 transistor, the first letter tells us what semiconductor material it is made of; the letter 'A' means germanium and 'B' means silicon. The second letter indicates the main use for the transistor; e.g., the letter 'C' indicates it can be used as an *audio frequency* amplifier. The letter 'S' would mean its main use is for *switching*; and a letter 'F' for *radio frequency* amplification.

Note that each transistor has three terminals, pins or leads. These terminals are called *collector* (c), *base* (b) and *emitter* (e). In the case of the TO3 transistor, the collector lead is its metal case. The popular transistors BC107 to BC108, 2N3053 and BFY51, have their metal cases connected internally to their collectors. The leads of a transistor are always identified by looking at its underside, not its top.

Note that the transistors shown are based on the semiconductor silicon. Germanium is still used for making some special types of transistors. If you come across an old transistor, especially one made before the early 1960s, it is most likely made of germanium. A germanium transistor can usually be identified by its long leads, at least four times longer than its case. Modern silicon transistors have leads about the same length as their cases.

## *Questions*

1  Do you know why earlier germanium transistors had such long leads?
2  Why do some transistors, e.g. the 2N3055 and BD121, have holes in their metal parts?

## 17.3  **Circuit symbols and internal structure for bipolar transistors**

There are two types of bipolar transistor: the *npn type* and the *pnp type*. The circuit symbols are shown in figure 17.2. The arrow within the symbol identifies the type: if it points from the base to the emitter, it is an npn type; if it points from the emitter to the base, it is a pnp type.

There are two p-n junctions in a bipolar transistor. The two arrangements of these junctions for the npn and pnp transistors are shown in figure 17.3. The base terminal is connected to a very thin layer of p-type or n-type material sandwiched

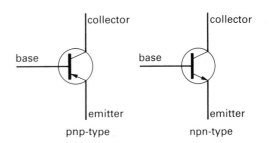

**Figure 17.2** Bipolar transistor symbols

between two layers of the opposite type of material.

Since the bipolar transistor comprises two p-n junctions, it is useful to think of the transistor as made up of two diodes connected together as shown in figure 17.3. If you bear this model in mind, it is easy to find out whether an unknown transistor is npn or pnp.

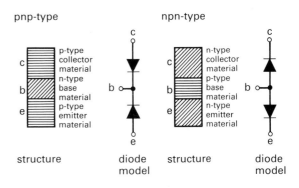

**Figure 17.3** Structure and diode model of a pnp and an npn transistor

## 17.4  *Experiment* C9

## **Identifying a transistor's leads**

Transistors must be connected correctly in a circuit otherwise they may be permanently damaged. The diagrams in figure 17.1 will help you identify the emitter, base and collector leads for commonly-used transistors. You should refer to the data supplied by the component distributers if you are in doubt about the connections on a transistor. But if you are about to use a transistor of unknown type, i.e. npn or pnp, the following simple test will give the answer. It will also allow you to test a transistor which you suspect is damaged.

The diagram in figure 17.4 assumes you are testing an npn transistor, e.g. the TO18, TO39 or TO5 outline. If the transistor is not marked, and you do not know the type, you will not be able to distinguish the collector terminal from the emitter terminal. However, you can still find out whether the transistor is npn or pnp. If the base terminal is not known, you can easily identify the base and the type of transistor it is.

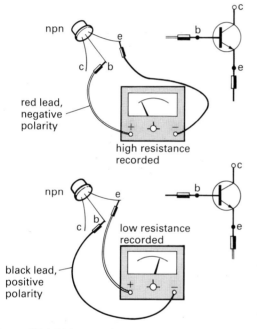

**Figure 17.4** Using an ohmmeter to test a transistor

Switch the multimeter to its 'ohms × 1' or 'ohms × 10' range. Attach the red lead to the base terminal and the other lead to the collector or emitter terminal of the transistor. Note the ohmmeter reading — not precisely, just roughly. In figure 17.4 the reading is shown high. The multimeter leads are then changed over and the multimeter should now show a low reading — about 50 ohms. If the transistor were a pnp type, the readings would have been reversed.

Now you should remember that the red lead has a negative polarity and the black lead a positive polarity when an analogue multimeter is switched to ohms. Thus if the transistor is an npn type, the internal diode between the base and collector, or base and emitter, is reverse biased by the first connection (high resistance) shown in figure 17.4, and forward-biased by the reverse connection (low resistance). For a pnp transistor, the internal diodes would be forward-biased by the first connection (low resistance) and reverse-biased by the second connection (high resistance).

The rule is: identify the polarity of the lead connected to the base terminal which gives the lower resistance between the base and collector or base and emitter. If it is the red lead (negative polarity), the base is n-type and the transistor is pnp; if it is the black lead (positive polarity), the base is p-type and the transistor is npn.

Clearly, it is easy to find the base lead of a transistor of unknown type. Just connect the black lead of the multimeter to one of the transistor leads, and connect the red lead to each of the other leads in turn.

If two low readings are found for one connection of the black lead, while each of the other two positions gives two high readings, then the transistor is an npn type, and the black lead is connected to the base.

If two high readings are found for one position of the black lead, while each of the other two positions gives two low readings, then the transistor is a pnp type, and the black lead is connected to the base.

If two low resistance readings, or two high resistance readings, are found when the leads of the multimeter are reversed, the transistor is damaged and should be thrown away.

It is not possible to use the ohmmeter in this way to distinguish between the emitter and collector leads; it is best to carry out a simple measurement of the d.c. current gain of the transistor to determine this, as explained in Section 19.2.

# 18 The Transistor as a Switch

## 18.1 *Experiment* C10

### Using a transistor as a switch

Figure 18.1 is a simple circuit which shows the way a transistor works as an electronic switch. Figure 18.2 shows how to wire up this circuit using terminal block.

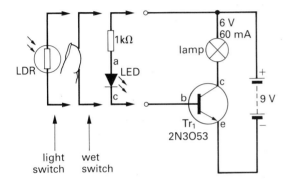

**Figure 18.1** Simple circuit to show the use of an npn transistor as a switch

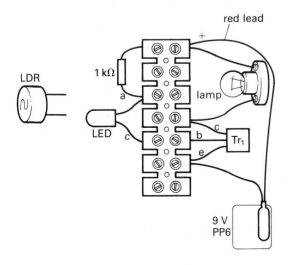

**Figure 18.2** The assembly of circuit 18.1 on terminal block

The light emitting diode (LED) shows when current flows into the base terminal of the transistor. The filament lamp shows when current flows into the collector terminal of the transistor. These two currents are shown in figure 18.3.

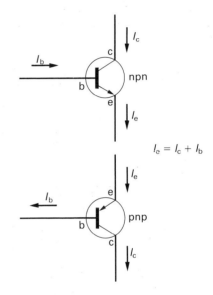

$$I_e = I_c + I_b$$

**Figure 18.3** Current flow into and out of an npn and a pnp transistor

This simple circuit shows the way a transistor is often used: a small current (about 5 mA) flowing through the LED into the base terminal of the transistor switches on a larger current (about 60 mA) which flows into the collector terminal of the transistor. The transistor is acting as an electronic switch: the lamp is switched on by a small current flowing into the base terminal.

Now remove the LED and put two bare wires in its place. Moisten the end of a finger and place the finger across these two wires. Again the lamp will light. Here you have a *wet-sensitive switch*.

Now remove the wires and replace them by a light dependent resistor (LDR). You can switch the lamp on by allowing light to fall on the LDR. You can switch the lamp off by covering the LDR. Here you have a *light-operated switch*.

These are simple electronic switches based on an npn transistor. A pnp transistor could have been used to do the same job but npn transistors are more widely used.

## 18.2   Current flow into and out of a bipolar transistor

Figure 18.3 shows the three currents which are controlled by a transistor. The current flowing into a transistor is equal to the current flowing out of the transistor. So whether the transistor is an npn or pnp type, we can write

$$I_c = I_e + I_b$$

However, since the base current is usually less than 1% of the collector current, we can see that $I_c = I_e$, approximately.

## Questions

**1** If $I_e = 10.1$ mA and $I_c = 10$ mA, what is the value of the base current?

**2** In a certain transistor, the base current is 0.5% of the emitter current. If the emitter current is 50 mA, what are the base and collector currents?

## 18.3 *Experiment* CII

## Measuring the switch-on voltage of a transistor

The circuit shown in figure 18.4 is a slight modification of that shown in figure 18.1. It is a practical circuit in which the transistor switches on the lamp when the

LDR is covered. The circuit can be readily assembled on terminal block as shown in figure 18.5.

The fixed-value resistor, $R_1$, and the LDR are connected in series and form a voltage divider (potential divider). The voltage across the LDR varies with its illumination. If it is covered, its resistance increases and the voltage across it rises. If light is allowed to reach it, its resistance decreases and the voltage across it falls. This action of a voltage divider is explained in Chapter 5, Book B. Note that the voltage across the LDR is also the voltage across the base-emitter terminals of the transistor.

Set a multimeter to its 1 V or 3 V d.c.

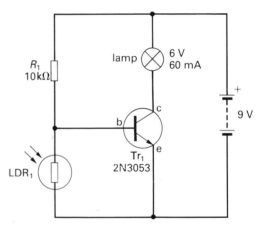

**Figure 18.4** Circuit for a dark-operated switch

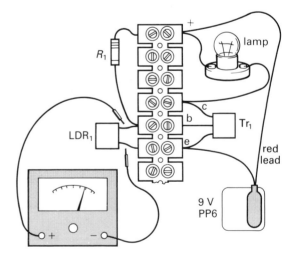

**Figure 18.5** Assembly on terminal block of the dark-operated switch

range and connect it across the LDR as shown. Allow light to reach the LDR so that the lamp goes off. Note the reading of the voltmeter — it should be less than half a volt. Now slowly cover the LDR and you will see the voltage rising. As it nears 0.6 V, the lamp will begin to light. Continue to cover the LDR until the lamp is at maximum brightness. What is the voltage across the LDR?

For a silicon bipolar transistor, this voltage should be between 0.6 V and 0.8 V and it is known as its switch-on voltage. A germanium transistor has a switch-on voltage of about 0.2 V.

## Remember

A silicon npn bipolar transistor switches on when its base-emitter voltage equals about 0.6 V.

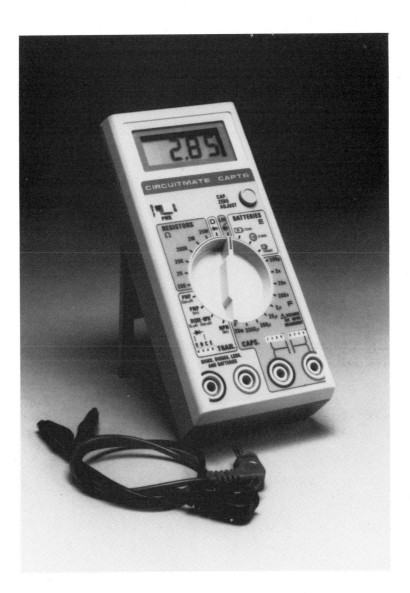

**Figure 18.6** The *Circuitmate*, a digital multimeter and component tester. It tests npn and pnp transistors as well as resistors, capacitors and diodes.    Courtesy: Beckman Industrial

# 19 The Current Gain of Bipolar Transistors

## 19.1 The meaning of current gain

Figure 18.3 shows the current flowing into and out of npn and pnp transistors. Experiment C10 in Section 18.1 showed that the base current is a small fraction of the collector and emitter currents. A small base current, $I_b$, is able to switch on a much larger collector current, $I_c$.

The ratio of the steady collector current to the steady base current is known as the *d.c. current gain* of the transistor. The symbol often used for d.c. current gain is the greek letter beta, $\beta$. But the symbol $h_{FE}$ is also used.
Therefore

$$\beta \text{ (or } h_{FE}) = \frac{\text{steady collector current}}{\text{steady base current}}$$

$$= \frac{I_c}{I_b}$$

Section 19.2 explains how to measure the d.c. current gain of a bipolar transistor.

## *Questions*

1 A circuit which used a ZTX300 transistor was able to switch a collector current of 180 mA when a base current of 5 mA flowed. What was the current gain of the transistor? What was the emitter current?
2 Under certain conditions the transistor BC108 requires a base current of 0.25 mA to make a collector current of 100 mA flow. What is the gain of the transistor under these conditions?

## 19.2 *Experiment* C12

### Measuring the d.c. current gain of a transistor

It is easy to use the simple transistor switching circuit of figure 18.4 to measure the current gain of the transistor. Figure 18.1 shows where ammeters (A) need to be connected to find the base current and the collector current. The circuit should be connected up on terminal block as shown in figure 18.5.

Switch a multimeter to its 100 mA (or nearest) range and connect it in series with the lamp as shown. With the battery connected to the circuit, cover the LDR so that the lamp switches on at its maximum brightness. Record this collector current, $I_c$.

Remove the milliammeter and reconnect the lamp in the circuit. Next connect the milliammeter in the base circuit as shown. Now cover up the LDR slowly and watch

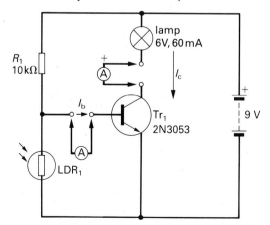

**Figure 19.1** Circuit for measuring the d.c. current gain of an npn transistor

the brightness of the lamp. As soon as the lamp reaches maximum brightness, measure the base current, $I_b$. This base current is much smaller than the collector current so you will need to switch the multimeter to its 3 mA (or lower) range to record it. If you have two multimeters it is possible to measure these two currents simultaneously.

Work out the ratio of the collector current to the base current. This value is the d.c. current gain of the transistor as explained in Section 19.1. Of course, a similar measurement would give the current gain of a pnp transistor.

Note that it is very important to measure the base current when the transistor just switches on the maximum current through the lamp. If you continue to cover the LDR, the base current will

continue to rise though the lamp remains at its maximum brightness. The collector current is said to *saturate* — see Section 25.2.

## 19.3 The electrical characteristics of some transistors

Bipolar transistors are chosen for a particular application on the basis of the electrical characteristics shown in figure 19.2. This table lists d.c. current gain, maximum collector current, and maximum power dissipation for a few common silicon transistors.

Note that the d.c. current gain does not have one precise value. For example, a particular BC107 transistor might have any current gain in the range 100 to 450. This is the manufacturing tolerance of this type of transistor. Fortunately, it is never necessary to know what the current gain of the transistor is exactly; most circuit designs will work with transistors in a wide gain-band. As a general rule, the circuits in *Basic Electronics* will work with any general-purpose transistor which has a gain of at least 100. Thus the BC182L, BC107 to BC109 and ZTX300 are all interchangeable in low-power transistor circuits.

In general, the higher the maximum collector current a transistor will switch, the lower the d.c. current gain of the transistor. Thus a power transistor, such as the 2N3055, can switch a collector current of 15 A, but its gain is only about 50. The low-power transistor, BC109, has a maximum collector current of only 100 mA but its gain may be as high as 900.

**Figure 19.2** The electrical characteristics of some common silicon transistors – see figure 17.1 for their outlines

| code | pnp/ npn | d.c. current gain | maximum collector current (mA) | maximum power (mW) |
|---|---|---|---|---|
| BC107 | npn | 100–450 | 100 | 360 |
| BC109 | npn | 200–800 | 100 | 360 |
| 2N2906 | pnp | 100–300 | 600 | 400 |
| 2N3704 | npn | 90–330 | 800 | 360 |
| BFY51 | npn | 40 | 1000 | 800 |
| ZTX300 | npn | 50–300 | 500 | 300 |
| 2N3053 | npn | 50–250 | 1000 | 800 |
| BC477 | pnp | 100–950 | 150 | 360 |
| TIP31 | npn | 10–60 | 3000 | 4000 |
| 2N3055 | npn | 20–70 | 15000 | 115 W |
| BC182 | npn | 100–480 | 200 | 800 |
| BC212 | pnp | 60–300 | 200 | 300 |

# 20 The Darlington pair Transistor

## 20.1 What it does

If you have built the circuit in figure 19.1, you may notice that the transistor does not switch on the lamp sharply. You can see the lamp brighten and fade when the transistor is switching on and off. This is because we are using a single transistor with a d.c. current gain of only a few hundred. What is needed is a circuit which switches the lamp on and off sharply at a particular level of illumination. A combination of two transistors, called a *Darlington pair*, will give us the 'snap-action' switch we want because it is designed to provide a very high current gain.

## 20.2 *Experiment* C13

### Making a Darlington pair

Use terminal block as shown in figure 20.2 to assemble the Darlington-pair circuit of figure 20.1. Note the use of the wire link in figure 20.2 to connect the emitter terminal of transistor $Tr_1$ to the base terminal of transistor $Tr_2$. This is an important connection in the Darlington pair for it shows that the current flowing out of the emitter terminal of $Tr_1$ forms the base current of $Tr_2$.

If you vary the amount of light reaching the LDR, the Darlington pair will switch the lamp on and off. Cover the LDR and the lamp will light; uncover it and the lamp will go out. You should notice that, compared with the single transistor switching circuit of figure 18.4, the Darlington pair is more sensitive; only a small change of illumination will switch the lamp on or off. This is now a snap-action circuit and practical applications for it are described in Chapter 21.

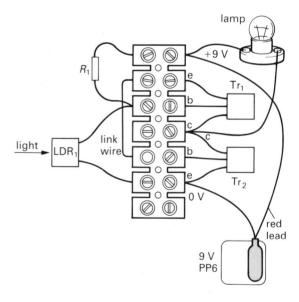

**Figure 20.2** Assembly of the Darlington pair using a terminal block

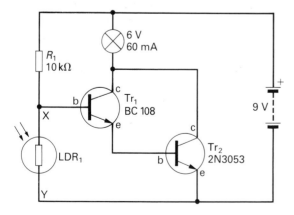

**Figure 20.1** Two transistors make a Darlington pair

## 20.3 How current flows in the Darlington pair circuit

Figure 20.3 shows two npn transistors connected together as in figure 20.1. One transistor switches more current through the lamp than the other. Which one?

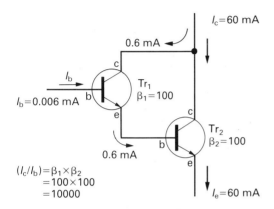

$(I_c/I_b) = \beta_1 \times \beta_2$
$= 100 \times 100$
$= 10000$

**Figure 20.3** The direction of current flow through a Darlington pair

You know that the emitter current of a transistor is about equal to its collector current (Section 18.2). Now the base current of $Tr_2$ is equal to the emitter current of $Tr_1$. Therefore the collector current of $Tr_1$ is very nearly equal to the base current of $Tr_2$. If $Tr_2$ has a current gain of 100, its base current (the collector current of $Tr_1$) is equal to 60 mA/100 or about 0.6 mA. About a hundred times more current flows through $Tr_2$ than through $Tr_1$. In practical circuits (see Chapter 21) the second transistor must be able to carry most of the current which flows through the lamp (or whatever device is being switched on and off).

If the current gain of $Tr_1$ is also 100, its base current is its collector current (about 0.6 mA) divided by its gain, i.e. 0.6 mA/ 100 = 0.006 mA or 6 μA. This is a very small current and it is the reason why the Darlington pair is so sensitive to changes of illumination of the LDR.

## 20.4 The d.c. current gain of a Darlington pair

As shown in figure 20.3, the Darlington pair can be regarded as a single transistor. The two collector terminals make one collector terminal; the base terminal of $Tr_1$ is the base terminal of the pair and the emitter terminal of $Tr_2$ is the emitter terminal of the pair.

The overall current gain of the Darlington pair is its collector current, $I_c$, divided by its base current, $I_b$. Let's assume that the current gain of each transistor is 100. The collector current equals 60 mA and the base current equals 6 μA, as was calculated above. Therefore the overall d.c. current gain of the Darlington pair equals 60 mA/6 μA = 10 000. Ten thousand is 100 × 100, i.e. the product of the two current gains. So this is how to calculate the d.c. current gain of a Darlington pair: simply multiply together the individual gains of the two transistors.

## Questions

1 A Darlington pair is made up of a BC109 transistor which has its emitter terminal connected to the base of a 2N3053 transistor. The current gain of the BC109 is 300 and of the 2N3053, 120. What is the overall current gain of the Darlington pair?

2 Use the list of electrical characteristics of transistors in figure 19.2 to design a Darlington pair which has a current gain of at least 5000, and which is able to switch a collector current of at least 2 A.

## 20.5 The base-emitter voltage required to switch on a Darlington pair

In a Darlington pair, both transistors have to be switched on since the emitter current of the first supplies the base current of the second. Now you know that the base-emitter voltage of an npn silicon transistor must be about 0.6 V to make current flow in its collector circuit. Thus the base voltage required to switch on a Darlington pair is $2 \times 0.6$ V = 1.2 V.

Check this voltage by connecting a multimeter switched to its 3 V d.c. range across the points X and Y in the circuit of figure 20.1. Cover the LDR until the lamp switches on and measure the steady base-emitter voltage. A value between 1.2 V and 1.6 V should be obtained.

### Remember

An npn Darlington pair made from two npn transistors switches on when its base-emitter voltage equals about 1.2 V.

## 20.6 Integrated circuit Darlington pairs

As you have seen, the Darlington pair gives a very high current gain. In other words, it requires a very small current into its base terminal to switch a much greater current in its collector circuit. This happens to be very useful in many electronic systems, especially computer systems, which do not have the ability to control lamps, relays and motors under their own power.

Microcomputer systems have eight (and sometimes 16) data lines for inputting and outputting data. In order to make use of each of these lines, Darlington pairs are available in integrated circuit packages. One such package is shown in figure 20.4. It is an 8-way Darlington driver so that it allows 8-bit microcomputers to control power-needy devices via its data lines. Note that each Darlington pair in this integrated circuit has input protection resistors to each base terminal, and output diodes to protect the transistors from back e.m.f. when the integrated circuit drives relays. When an input signal is *high*, the corresponding collector terminal at the output becomes *low* and current flows from the positive supply into the integrated circuit. This Darlington driver is used in Project Module D2.

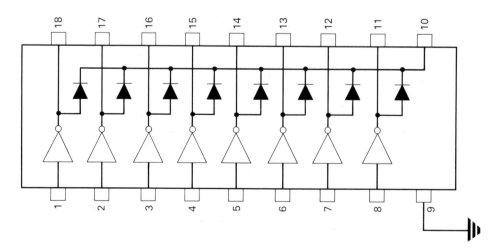

**Figure 20.4** Darlington pairs in an integrated circuit package, e.g. type ULN 2803 A

# 21 Practical Transistor Switching Circuits

## 21.1 The electronic system of a switching circuit

The single transistor (figure 18.1) and Darlington pair (figure 20.1) switching circuits are each made up of three main building blocks which form the electronic system shown in figure 21.1.

A variety of sensors may be used which respond to such things as light, heat, magnetism and touch. In the case of bipolar transistors, the amplifier amplifies current. The device which is switched on and off may be a lamp, electric motor or a relay, which in turn might control a lamp or motor.

## 21.2 Practical circuits

Figure 21.2 shows how a variety of sensors might be connected to a Darlington pair amplifier. You could use a single transistor but it would not give such a sensitive response. Resistor $R_2$ prevents too much current flowing through the base-emitter junctions of $Tr_1$ and $Tr_2$ should point X be connected accidentally to the positive supply.

**(a) Light switch** The operation of this circuit has been discussed in Section 20.2. If the LDR is covered, its resistance rises. When the voltage at point X reaches about

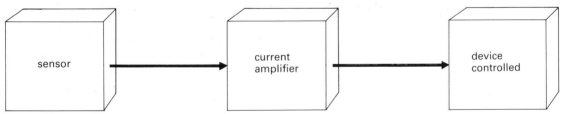

**Figure 21.1** Electronic system of the transistor switching circuit

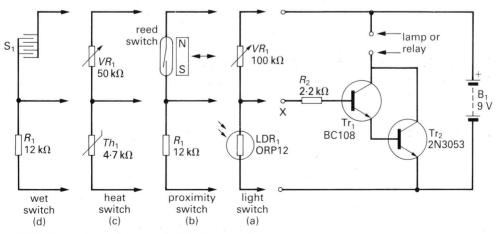

**Figure 21.2** Practical circuits using the Darlington pair

1.2 V, the Darlington pair switches on and lights the lamp or energises the relay. The relay can be used to switch on electric motors and other devices which require a larger current than $Tr_2$ can supply.

## Question

**1** What does the circuit do if the LDR and variable resistor are interchanged?

**(b) Proximity switch** If the magnet is moved up to the reed switch, its contacts close and the transistors switch on. This circuit can be used to switch on a relay or lamp when a reed switch passes a magnet. In this circuit the relay carries a small current. This is unlike the simple circuits described in Book A, Chapter 5, where all the current to a motor or lamp passes through the reed contacts.

## Question

**2** What does the circuit do if the reed switch and resistor $R_1$ are interchanged? Note that the value of $R_1$ is not critical so use any value between 1 k$\Omega$ and 47 k$\Omega$.

**(c) Heat switch** This circuit acts as a simple thermostat. As the thermistor is heated its resistance falls. When the voltage at X has fallen below about 1.2 V, the lamp switches off. A relay in place of the lamp can be used to switch an electrically heated element on and off. The variable resistor is used to set the circuit to switch off the relay at a pre-determined temperature. For example, a fish tank could be maintained at a temperature of 45°C. A 50 k$\Omega$ variable resistor is suitable for a thermistor which has a value at 25°C of about 4.7 k$\Omega$.

## Question

**3** Is there any use for the circuit if you interchange the thermistor and variable resistor?

**(d) 'Wet' switch** The sensor $S_1$ is made up of two sets of conductors, e.g. the copper strips on stripboard. If the gaps separating the conductors are dry, the resistance of the sensor is high and the voltage at point X is very small but current flows across the gaps if the sensor is damp. Its resistance decreases and the voltage at X rises. When it reaches about 1.2 V, the transistors switch on. A slightly redesigned circuit could be used to sound an alarm when the soil in a plant pot becomes dry. You would have to experiment with the value of $R_1$ to obtain the switching point you wanted or use a 10 k$\Omega$ variable resistor.

21.3 **Alternative two-transistor switching circuits**

The circuit shown in figure 21.3 uses two npn transistors in a light-sensitive switching circuit. It is not a Darlington pair but a two-stage transistor amplifier in which $Tr_1$ can switch on before $Tr_2$. Remember that in a Darlington pair, both transistors switch on together since the emitter current of the first delivers current to the base of the second (see Section 20.3).

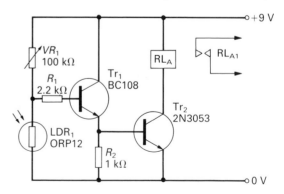

**Figure 21.3** An alternative two-transistor switch

However, in the two-stage amplifier shown in figure 21.3, current flows out of the emitter of $Tr_1$ through resistor $R_2$ before transistor $Tr_2$ switches on. Only when the voltage across $R_2$ has risen to 0.6 V does $Tr_2$ switch on to make current flow through the lamp.

This is the way the circuit works: suppose the voltage at point X is less than 0.6 V, i.e. the LDR is lit so its resistance is low compared with the value of resistor $R_1$. $Tr_1$ is off, no current flows through resistor $R_2$ so the voltage at point Y is 0 V. Transistor $Tr_2$ is also switched off and the lamp is off.

When the voltage across the base-emitter junction of $Tr_1$ reaches 0.6 V, this transistor switches on and the lamp lights. Figure 21.3 shows the currents and voltages on the circuits when the lamp is on. Note that the voltage at the base of the $Tr_1$ equals the sum of the voltages across $R_2$ and the base–emitter junction of $Tr_1$, i.e. 1.2 V.

It is easy to work out a suitable value for $R_2$ if we know the d.c. current gain of $Tr_2$ — suppose it is 100. When the lamp switches on fully, a current of 60 mA flows through it. Thus the base current of $Tr_2$ is 60 mA/100 = 0.6 mA.

As a general rule the current flowing through $R_2$ when $Tr_2$ switches on should be about ten times the base current of $Tr_2$, i.e. 6 mA. The value of $R_2$ should be 0.6 V/6 mA = 100 $\Omega$. The circuit will work well with resistors in the range 100 $\Omega$ to 1 k$\Omega$ without a noticeable change in its sensitivity.

## Question

1  Figure 21.4 shows an alternative two-stage amplifier. How does this circuit work when the LDR is covered and uncovered?

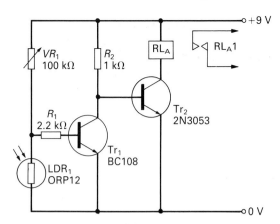

**Figure 21.4**  A two-transistor switch

## 21.4  Transistorised time-delay circuit

A capacitor can be used in the Darlington pair switch in figure 21.2, and the two-transistor switch of figure 21.3 and 21.4, to produce time delays of up to about five minutes. Figure 21.5 shows a practical circuit using a Darlington pair.

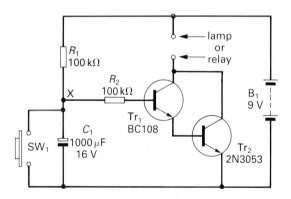

**Figure 21.5**  Time delay circuit

After releasing the push switch, $SW_1$, the capacitor, $C_1$, begins to charge through resistor $R_1$ as explained in Book B, Chapter 10. The time constant, $T$, determines the rate at which the voltage at point X rises ($T = C_1 R_1$). In this circuit, the time constant $T = 1000 \times 10^{-6} \times 100 \times 10^3 = 100$ s. This is the time for the voltage to rise to about 2/3 of the supply voltage. If this is 9 V, it takes 100 s to reach 6 V.

However, since point X is connected, via resistor $R_2$, to the base of $Tr_1$, the voltage never reaches 6 V. Instead it rises to only about 1.2 V (ignoring the voltage across the low-value base resistor, $R_2$) when the Darlington pair switches on. The variation of voltage at point X is shown in figure 21.6.

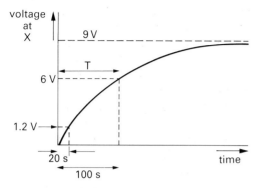

**Figure 21.6** Variation of voltage at point X in figure 21.5

So what is the time delay before the transistors switch on? If we assume the voltage on plate X of the capacitor varies linearly, i.e. the graph up to 100 s is a straight line, the answer is easy. Since 1.2 V is one fifth of 6 V, the time delay is $(100/5) = 20$ s. If $R_1$ is replaced by a variable resistor, the time delay may be adjusted: a 1 MΩ variable resistor will give time delays up to 200 s.

However, for two practical reasons long time delays are not possible with this circuit. The main problem is that electrolytic capacitors suffer from a leakage current. Therefore high values of $R_1$ are not possible because the current flowing through $R_1$ is so small it simply leaks away through the capacitor. Practical limits to the values of $R_1$ and $C_1$ are about 1 MΩ and 2000 μF, respectively. That means a maximum time delay of about 400 s, i.e. 6 minutes. However, it is worth experimenting with values of $R_1$ and $C_1$. A second practical limitation is that the Darlington pair does need a small input current to work. $R_1$ should not have too high a value otherwise this current will be too low.

## 21.5  Power dissipation in a transistor

Figure 21.7 shows a single transistor used to switch a lamp on and off. When the reed switch is closed, the lamp lights; when it is open the lamp goes off. Resistor $R_1$ protects the transistor from damage when the reed switch is closed. Its value is determined by the gain of the transistor, the maximum collector current which flows through the lamp, and the power supply voltage.

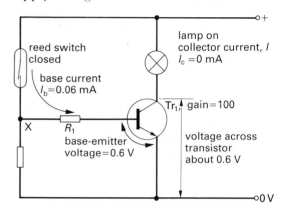

**Figure 21.7** Calculating the power dissipation in a transistor

It is easy to calculate the maximum value of resistor $R_1$ to ensure that the transistor is saturated. If the transistor has a current gain of 100, the base current $I$ is given by

$I = 60$ mA/100 = 0.6 mA.

When the reed switch is closed, the voltage at point X is 6 V. The potential difference across $R_1$ is $(6 - 0.6)$ V = 5.4 V. Therefore the value of $R_1$ is given by

$R_1 = 5.4$ V/0.6 mA = 9 kΩ.

It is better to use a lower value than this to ensure that the transistor is fully switched on and in case the gain of the transistor is lower than 100. A value of 2.2 kΩ has been used in the transistor switching circuits described in this book.

If a Darlington pair circuit is used, the base resistor $R_1$ can have a much higher value since much less current flows into the base of the first transistor.

When the transistor is fully switched on, almost all the supply voltage occurs across the lamp. In fact only about 0.6 V occurs across the transistor when it is fully switched on. The heat produced in the transistor at saturation is given by

current through transistor
× voltage across transistor
= 60 mA × 0.6 V = 36 mW.

This is a small amount of heat which most low-power transistors can dissipate without becoming too warm. However, if the transistor is not fully saturated, more heat is produced in it. For example, if the current through the transistor is 30 mA the voltage across it might be 5 V. The power dissipation in the transistor is now 5 V × 30 mA = 150 mW and this can cause the transistor to become warm.

Unless a heatsink is fitted to the transistor it may become overheated and fail.

## *Questions*

1 Calculate the maximum value of the resistor in the base circuit of a Darlington-pair transistor switch to make sure that the transistors are fully switched on. Assume the overall gain of the transistors is 5000, the collector current which flows is 100 mA and the supply voltage is 10 V.
2 If the voltage across a transistor which is not fully switched on is 6 V and the current which flows through it is 20 mA, what is the power dissipation in the transistor?

## 21.6  **Transistor heatsinks**

The heat generated by current flowing between the collector and emitter

junctions of a transistor causes its temperature to rise. This heat must be conducted away from the transistor otherwise the temperature rise may be high enough to irreparably damage the p-n junctions inside the transistor.

For low-current transistors such as the BC108, ZTX300 and BC182L the heat generated does not usually exceed 100 mW. This small amount of power (100 millijoules per second) easily escapes by conduction along the leads and by convection to the surrounding air.

Transistors such as the 2N3053, TIP32 and 2N3055 dissipate more heat since the current passing through them is higher. These transistors must be fitted with a heatsink to help get rid of the heat produced within them. Figure 21.8 shows three heatsinks used for three common shapes of transistors. The colour of the heatsink is usually black for this improves the rate at which heat leaves it by *radiation*. Most heatsinks are corrugated or finned which helps to dissipate heat to surrounding air by *convection* since their surface area is greater.

The heatsink for the TO5 and TO39 type of transistor (e.g. 2N3053 and BFY51) simply clips over the metal can

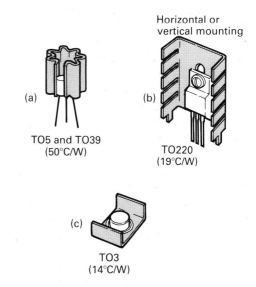

**Figure 21.8** Three types of transistor heatsink (see figure 21.1 for outlines).

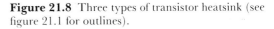

(figure 21.8(a)). This allows the transistor to operate safely at its maximum power dissipation (about 800 mW for the 2N3053).

The TIP31A and TIP121 transistors (and many others) have a metal plate on them drilled with a hole so that they can be bolted firmly to a heatsink of the type shown in figure 21.8(b). Similarly, the TO3 type of transistor (e.g. the 2N3055) is encased in a metal package drilled with two holes to allow it to be bolted to the type of heatsink shown in figure 21.8(c).

## ▽ 21.7   **Choosing and using a heatsink**

If you look at heatsinks in a catalogue, you will see that they are rated by a quantity called *thermal resistance*. Thermal resistance is measured in units of degrees centigrade per watt, or °C/W, and has the symbol $\theta$ or $R_{th}$. Large heatsinks for TO3-type high-power transistors have a low thermal resistance of about 0.5°C/W. Small heatsinks for low-power TO5-type transistors have a high thermal resistance of about 50°C/W (see figure 21.8).

Thus thermal resistance is the yardstick for comparing heatsinks. The lower the value of the thermal resistance, the better the heatsink is at getting heat away from the transistor. A simple equation is used to find the thermal resistance of a heatsink required for a particular transistor:

$R = (T_h - T_a)/W$

where $T_h$ is the temperature of the heatsink

$T_a$ is the ambient temperature

$W$ is the power required to be dissipated from the transistor mounted on the heatsink

Suppose a transistor is to dissipate 15 W and you want to know the thermal resistance of the heatsink you are to use. The manufacturer's data also gives the maximum safe temperature for the case of their transistor, usually 125°C. If proper precautions are taken (see below), the temperature of the heatsink will be the same as the temperature of the case of the transistor. Therefore the above equation tells us that the thermal resistance required is

$$\frac{125 - 20}{15} = 7°C/W.$$

It is as well to be on the safe side, so choose a heatsink which has a *lower* thermal resistance than this, e.g. 4°C/W.

The above calculation assumes that you have made such good thermal contact between the case of the transistor and its heatsink that the temperature of the heatsink is the same as that of the transistor. Make sure they are by mounting the heatsink as follows:

(i) Make sure that the surface of the heatsink where the transistor is to be mounted is absolutely flat and clean.

(ii) Remove any raised burrs on the metal surfaces, particularly round the fixing holes.

(iii) Spread a thin layer of 'heatsink compound' on this flat surface.

(iv) Finally, bolt the transistor to the heatsink using fixing screws, often supplied as part of the transistor's mounting kit.

The mounting kit supplied for power transistors generally includes a thin mica washer for fixing between the heatsink and transistor. This washer has a low thermal resistance and is used when it is necessary to electrically isolate the transistor from the heatsink, e.g. where the heatsink is bolted to an earthed metal case (usually at 0 V). Most power transistors have their collector terminals internally connected to the metal case or tab of the transistor. So without the washer the collector would be connected to 0 V with disastrous results. The washer, and the insulating bushes provided, should also be coated with
△ heatsink compound.

# 22 Why a Transistor Amplifies

## ▽ 22.1 Transistor switched off

The way a bipolar transistor amplifies current can only be explained in terms of the flow of electrons and holes through it. Chapters 15 and 16 introduced the properties of p-type and n-type materials and of the p-n junction. Figure 22.1 shows the distribution of electrons and holes in an npn transistor when it is switched off. As in the practical circuits described in Chapters 19 and 20, the collector is made positive relative to the emitter using the external voltage, $V_c$. The base is connected to 0 V so that there is no voltage across the base-emitter junction. Since the p-n junction formed by the collector and emitter regions is reverse-biased, electrons and holes move away from the collector-base junction. A depletion region is formed and this junction is reverse-biased by the external voltage $V_c$. Therefore current cannot flow in the collector circuit because of the very high resistance of this △ depletion region.

## ▽ 22.2 Transistor switched on

Figure 22.2 overleaf shows the flow of electrons and holes through the transistor when the base-emitter junction is forward-biased by an external source of voltage, $V_b$. A large number of electrons now move easily from the emitter to the base. A few of these electrons combine with holes in the base, but not many because the base region is lightly doped and very thin compared with the collector and emitter regions.

Because the base region is lightly doped, about 0.2 to 2% of the electrons moving from the emitter to the base combine with holes in the base. In other words, if the electron current from the emitter to the base is 100 mA, between about 0.2 mA and 2 mA of this current is 'lost' by recombination with holes in the base region. To make up this 'loss' of holes, a small electron current, the base current $I_b$, flows out of the base.

It is because the base region is so thin that most of the rapidly moving electrons

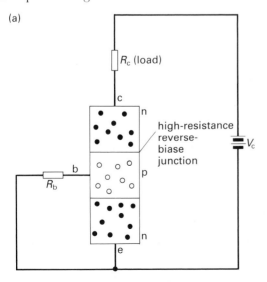

(a)

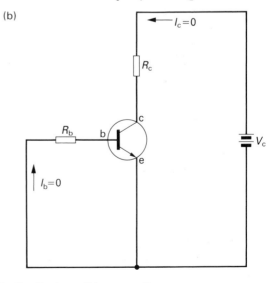

(b)

**Figure 22.1** Transistor switched off  (a) electron and hole distribution  (b) current flow

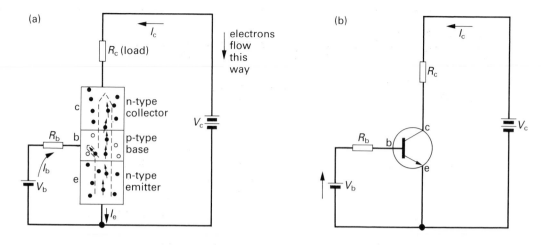

**Figure 22.2** Transistor switched on   (a)  electron and hole flow   (b)  current flow

which reach the base from the emitter are swept across the reverse-biased collector-base p-n junction and form the collector current $I_c$.

In effect the transistor has amplified a small base current to produce a larger collector current. If 0.2% of the electrons reaching the base from the emitter combine with holes in the base, and the emitter current is 100 mA, the base current is 0.2 mA and the current gain, $\beta$, of the transistor is given by:

$$\beta = \frac{\text{collector current}}{\text{base current}}$$

$$= \frac{I_c}{I_b}$$

$$= \frac{\text{emitter current} - \text{base current}}{\text{base current}}$$

$$= \frac{(100 - 0.2)}{0.2}$$

$$= 500$$

## Questions

**1** What is the current gain of a transistor if 2% of the electrons reaching the base from the emitter combine in the base region?

**2** An npn transistor is made so that 1% of the electrons reaching the base from the emitter combine with holes in the base. If the collector current is 60 mA, calculate the base current and the current gain of the transistor?

In an npn transistor, electrons are the majority charge carriers since the emitter and collector are made of n-type material. Holes are minority charge carriers — a very small flow of holes occurs from the base to the emitter region and is not shown on the above diagrams.
In a pnp transistor, holes are the majority charge carriers as shown in figure 22.3. Electrons are the minority charge carriers and come from the lightly doped n-type base region.

$\triangle$

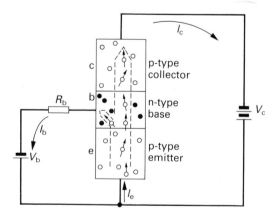

**Figure 22.3**  The flow of holes through a pnp transistor

# 23 Bipolar Transistor Characteristics

▽ 23.1 **Introduction**

The characteristics of a transistor are graphs which show the relationships between the various currents and voltages when the transistor is used as a switch or an amplifier. The graphs enable the circuit designer to see how best to use the transistor.

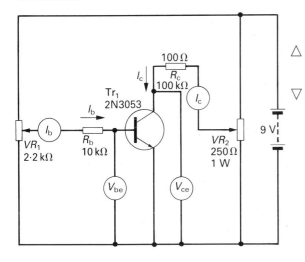

**Figure 23.1** A practical circuit for obtaining transistor characteristics

Three graphs showing these characteristics for an npn transistor can be obtained with the circuit shown in figure 23.1. The circuit enables four quantities to be measured:

$I_b$: the base current
$I_c$: the collector current
$V_{be}$: the base-emitter voltage
$V_{ce}$: the collector-emitter voltage

Resistor $R_b$ protects the transistor from excessive base currents. Resistor $R_c$ is the load resistor, i.e. a resistor through which the transistor drives current. In this circuit, the load resistor might be a 6 V, 60 mA lamp, or a relay, for example.

The three characteristics which can be obtained with this circuit are:

input characteristic
output characteristic
transfer characteristic

Note that, like the practical circuits in Chapter 21, this circuit uses the transistor connected as a common-emitter amplifier. That is, the input circuit and the output circuit have a common connection with the emitter terminal of the transistor.

▽ 23.2 **The input (or base) characteristic**

The input characteristic shows the relationship between the base current, $I_b$, and the base-emitter voltage, $V_{be}$. It is obtained by keeping the collector-emitter voltage, $V_{ce}$, constant and measuring $V_{be}$ for various values of $I_b$.

A typical graph for a silicon transistor is shown in figure 23.2 (overleaf). The main point to notice from this graph is that $I_b$ is negligibly small unless $V_{be}$ exceeds about 0.7 V. This critical voltage for a silicon transistor is generally known as the switch-on voltage: only when base current begins to flow does the transistor amplify a collector current.

This graph enables the input resistance of the transistor to be measured.

$$\text{input resistance, } R_i = \frac{\text{change in } V_{be}}{\text{change in } I_b}$$

Its value is not constant since the graph is curved, but it has a value between about 1 kΩ and 5 kΩ.

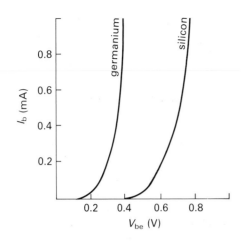

**Figure 23.2** Input characteristics of silicon and germanium transistors

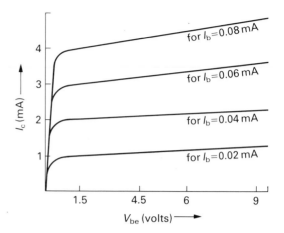

**Figure 23.3** Output characteristics of a bipolar transistor

Figure 23.2 also shows that an old-fashioned germanium transistor has a lower switch-on voltage of about 0.2 V.

▽ 23.3 **The output (or collector) characteristic**

The output characteristic shows the relationship between the collector current, $I_c$, and the collector-emitter voltage, $V_{ce}$. It is obtained by first fixing a low value base current, e.g. 10 μA, using potentiometer $VR_1$. Then the potentiometer, $VR_2$, is adjusted to give different values of $V_{ce}$ and corresponding values of $I_c$ can be measured. A family of output graphs is obtained for different values of $I_b$ as shown in figure 23.3.

As you can see, the curves are almost parallel to the $V_{ce}$ axis. This means that a large change in the collector-emitter voltage produces a very small change in the collector current (for a constant base current, of course). This means that the transistor behaves as though it is a large-value resistor. The output resistance, $r_o$, of the transistor is fairly high and is given by

$$\text{output resistance, } r_o = \frac{\text{change in } V_{ce}}{\text{change in } I_b}$$

Its value is between 10 kΩ and 50 kΩ.

This behaviour gives us the origin of the word 'transistor'. It comes from 'trans-resistor' which refers to the fact that a small current through the low resistance base-emitter junction switches a large current through the base-collector junction. This is the secret of a transistor operating as an amplifier: a small voltage change across the base-emitter junction produces a large voltage change across the base-collector junction. Chapter 24 describes this effect in more detail.

▽ 23.4 **The transfer characteristic**

If $V_{ce}$ is kept to a fixed value, say 6 V, using $VR_2$, and $VR_1$ is varied to change $I_b$, the graph shown in figure 23.4 is obtained. The graph is almost a straight line indicating that the $I_c$ is proportional to $I_b$ and amplified by a factor called the a.c. current gain $h_{FE}$:

$$h_{FE} = \frac{\text{change in } I_c}{\text{change in } I_b}$$

The d.c. current gain, which was discussed in Chapter 19, is measured for just one value of collector current, usually just

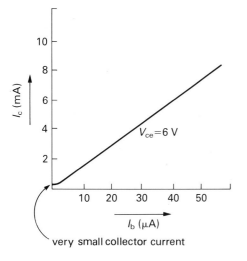

$V_{ce} = 6$ V

$I_c$ (mA)

$I_b$ (μA)

very small collector current

**Figure 23.4** Transfer characteristics of a bipolar transistor

when it reaches its maximum, or saturation, value. There is little difference between the a.c. and d.c. current gains.

Note that when the base current is zero there is almost zero collector current. A very small leakage current flows but it is

less than 0.01 μA for a silicon transistor, and rather higher, about 2 μA, for a germanium transistor. This current is caused by minority charge carriers (holes in an npn transistor) which flow from the collector to the base junction in the opposite direction to the flow of majority charge carriers (electrons in an npn transistor). A temperature increase causes the leakage current to increase, but it is so small for silicon transistors that it can be ignored. However, this is not so for germanium transistors, but these are hardly ever used nowadays.

## Remember

For a silicon transistor:
(i)  the collector current is zero until a base current flows;
(ii)  the base current is zero until the base-emitter voltage exceeds about 0.7 V
(iii)  the base-emitter voltage remains constant for a wide range of base current.

△

# 24 The Load Line for a Bipolar Transistor

## ▽ 24.1 A transistor voltage amplifier

Not only are transistors used as switches (see Chapters 18 and 20), but they are also used to amplify small varying voltages. For example, the circuit shown in figure 24.1 enables you to 'listen in' to the 'sounds of light'. The circuit will amplify the varying light intensity of a mains-operated lamp, or the light from a light transmitter in an optical communications system. (See Project Module B6 for a suitable light sensor.)

The photodiode, $D_1$, and resistor $R_1$ convert variations in light intensity into small changes of voltage at point X. These changes are passed to the base of the transistor via capacitor $C$, (known as a coupling capacitor). The transistor amplifies the small variations in base current to produce a larger current change through the load resistor, $R_c$. These amplified current variations produce large voltage variations at point Y which are passed into the earpiece via a second coupling capacitor, $C_2$. Resistor $R_b$ ensures that a small base current flows into the transistor. The value of this resistor is
△ selected as explained in Section 24.4.

## ▽ 24.2 The load-line

It is important to select the value of $R_b$ in the transistor amplifier of figure 24.1 so that there is no distortion of the amplified signal. By drawing a load-line on the output characteristics of the transistor (Chapter 23), the circuit designer can select a value for $R_b$ to make sure that distortion is minimised.

Figure 24.2 shows the currents and voltages in the output circuit of a transistor. The supply voltage, $V_c$, is shared between the transistor and the load resistor $R_c$. Therefore

$$V_c = V_1 + V_{ce} \text{ or } V_{ce} = V_c - V_1$$

However, $V_1 = I_c \times R_c$

so that $\quad V_{ce} = V_c - V_1 = V_c - I_c \times R_c$

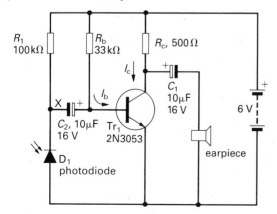

**Figure 24.1** A single transistor voltage amplifier

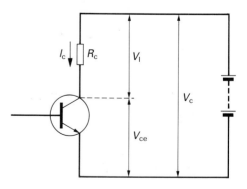

**Figure 24.2** Voltages acting at the output of a transistor

The equation $V_{ce} = V_c - I_c R_c$ represents the equation of a straight line, since, if the pairs of points $I_c$ and $V_{ce}$ are plotted, they lie on a straight line. This line is shown in figure 24.3 for the case of $R_c = 500$ $\Omega$ and $V_c = 6$ V, and is known as the *load-line* since it indicates the operating conditions of the transistor when it is driving current into a resistive load in the collector circuit. More accurately, this line is known as the *500 $\Omega$ load-line*, since it represents the relationship between $V_{ce}$ and $I_c$ for a load-resistance $R_c$ of 500 $\Omega$. Note the following points from figure 24.3.

**(a)** When $V_{ce} = 0$, $V_c = I_c R_c$. Therefore, $I_c = 6$ V/0.5 k$\Omega$ = 12 mA. Actually $V_{ce}$ is never quite zero, but 12 mA will be taken as the maximum collector current, known as *saturation* of the transistor.

**(b)** When $V_{ce} = 6$ V, $I_c R_c = 0$, or $I_c = 0$. This is the condition known as *cut-off*, when no current flows through the transistor.

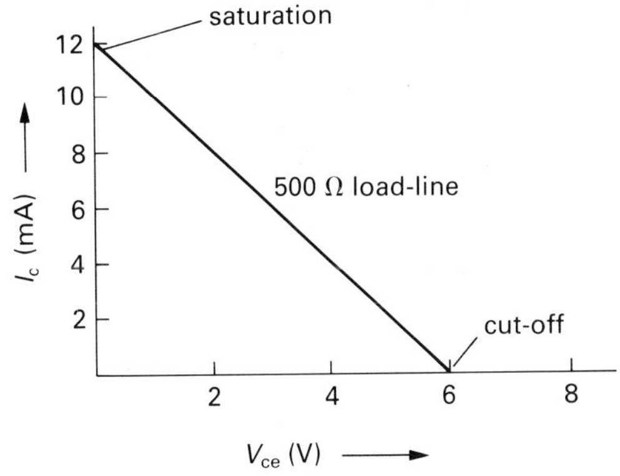

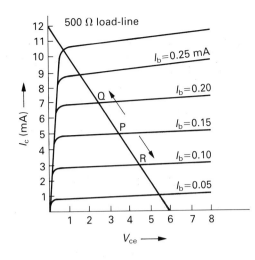

**Figure 24.4** A load-line superimposed on a set of transistor output characteristics

Figure 24.3 The straight line graph of a load-line

## ▽ 24.3 The transistor operating point

The transistor operating point must lie on the intersection of the load-line and the characteristic curve of the transistor. For instance, suppose the d.c. operating point lies at the point P on the $I_c/V_{ce}$

characteristics shown in figure 24.4. At this point, note that $I_c = 5$ mA, $I_b = 0.15$ mA, and $V_{ce} = 3.5$ V.

Now suppose that the base current is increased to 0.20 mA by an increase in the voltage applied across the base-emitter junction. Since the transistor must always operate on the load-line as well as on the characteristic curves, the d.c. operating point is now at Q; and if the base current falls to 0.1 mA, caused by a decrease in the voltage across the base-emitter junction, the operating point moves to R. Note, once again, that the transistor operates as a *current-amplifier*: a small change in $I_b$ gives rise to a large change in $I_c$. The a.c. *current gain* ($h_{FE}$) is the *change* in $I_c$ brought about by a *change* in $I_b$; that is

$$h_{FE} = \frac{\text{change in collector current}}{\text{change in base current}}$$

This can be calculated from figure 24.4. Base current change between the points Q and R = 0.1 mA; collector current change between the points Q and R = 4 mA; therefore $h_{FE} = 4/0.1 = 40$.

Compare this with the d.c. current gain at P, say. At P,

$$h_{FE} = \frac{\text{steady collector current}}{\text{steady base current}} = \frac{5}{0.15} = 33$$

The voltage gain, $A$, is given by

$$A = \frac{\text{change in output voltage}}{\text{change in input voltage}}$$

$$= \frac{\text{change in } V_{ce}}{\text{change in } V_{be}}.$$

We need both the output and the input characteristics of the transistor to find $A$. From the load-line in figure 24.4, we see that $V_{ce}$ changes by 2 V as the operating point changes between R and Q. This corresponds to a change in base current of $(0.20 - 0.10) = 0.1$ mA. Now assume that this silicon transistor has the input characteristic shown in figure 23.2. A change in $I_b$ of 0.1 mA is brought about by a change in $V_{be}$ of about 0.04 V. So the △ voltage gain is the ratio 2 V/0.04 V = 50.

▽ 24.4   **Choosing a value for the base resistor**

The above calculations show that an amplifier with a voltage gain of 50 can be built using a transistor with a current gain of 33. Suppose this transistor is used in the audio amplifier circuit shown in figure 24.1. We have decided on a load resistor, $R_c$, of 500 Ω. What we must now do is to choose a value for the base resistor $R_b$

which supplies the steady base current, $I_b$, of 0.15 mA. This current must set the transistor at its operating point, P, in figure 24.4. That is, $R_b$ must be chosen so that $I_b = 0.15$ mA. The voltage across resistor $R_b$ is equal to $(6 - 0.7)$ V = 5.3 V since the base-emitter voltage is equal to 0.7 V. Thus the value of $R_b$ is equal to (5.3 V/0.15 mA) = 33 kΩ, approximately.

So our design is complete. But $R_b$ must be chosen with care. If we choose too low a value for $R_b$, the base current will be too high and we shall shift our operating point towards Q; and if $R_b$ is too high, we shall move the transistor's operating point towards point R. Either way we will run the risk of distorting the amplified output waveform. Point P is chosen so that, in the absence of an input signal, the collector-emitter voltage is set at about 3 V, i.e. half the supply voltage. In this condition, the output voltage can swing by about 2 V each way without distortion. Any movement of the operating point towards Q or R would give rise to the type of distortion shown in figure 24.5, in which one half of the waveform is *clipped*, i.e. flattened. Remember, too, that even if the operating point is at point P, too large an input signal will cause clipping of both △ halves of the waveform.

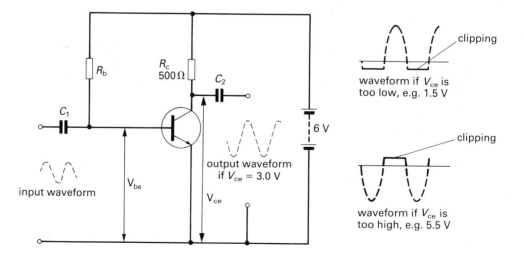

**Figure 24.5**  The output waveform is distorted if the base resistor does not have the right value

# 25 Negative and Positive Feedback

▽ 25.1 ## The transistor amplifier and negative feedback

When there is no input signal to the single transistor amplifier (figure 24.1), the voltage, $V_{ce}$, at the collector of the transistor should be at half the supply voltage, i.e. $V_{ce}$ must equal $V_c/2$ to get the maximum undistorted swing in the output voltage. This voltage is set by the value of the collector current flowing through $R_c$. If $I_c$ increases, $V_{ce}$ decreases, and vice versa.

There are two reasons why $I_c$ may change. The first reason is that transistors of the same type, e.g. 2N3053, usually have different current gains. So although $V_{ce} = V_c/2$ may be true for one transistor, a replacement transistor having a different gain would raise or lower $V_{ce}$ and make distortion more likely.

The second reason is that when transistors become warm (usually of their own making), $I_c$ increases due to the production of more electrons and holes in the transistor. This increased current causes more heating and hence more current, a situation known as *thermal runaway*. If unchecked, thermal runaway may destroy the transistor.

It is possible to stabilise the operation of the transistor amplifier to make sure that $V_{ce} = V_c/2$ even though $I_c$ may change. The trick used in circuit design is called *negative feedback* and is shown at work in the modified single transistor amplifier shown in figure 25.1. Note that the base resistor, $R_b$, is now connected between the collector and the base of the transistor (compare this with figure 24.1). This circuit is analysed as follows:

$$V_c = I_c \times R_c + V_{ce}$$

and $$V_{ce} = I_b \times R_b + V_{be}$$

The first equation shows that if $I_c$ tries to increase for any reason, $V_{ce}$ decreases.

From the second equation it therefore follows that $I_b$ must decrease ($V_{be}$ and $R_b$ are fixed). And a decrease in the value of $I_b$ must produce a decrease in the value of $I_c$, maintaining it at its original value.

What should be the value of $R_b$ in this case? Suppose we take the values of $V_{ce}$, $I_c$ and $I_b$ from the load-line in figure 24.4. At point P

$$I_c = 5 \text{ mA}; I_b = 0.15 \text{ mA}; V_{ce} = 3.5 \text{ V}$$

From the second equation above

$$R_b = \frac{(V_{ce} - V_{be})}{I_b} = \frac{(3.5 - 0.7) \text{ V}}{0.15 \text{ mA}} = 19 \text{ k}\Omega$$

You would then choose a resistor of preferred value 18 k$\Omega$. This value is about half that used in the non-stabilised amplifier (figure 24.1).

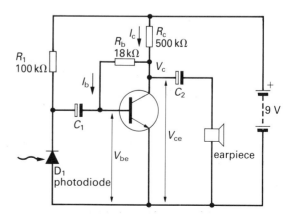

**Figure 25.1** Stabilising the operating point of a transistor using negative feedback

The stabilisation is effective because the circuit is designed to make use of negative feedback. This is how negative feedback works in this circuit: any unwanted increase in the output (i.e. an increase in $I_c$), causes a signal to be sent back to the input (via $R_b$) to reduce the current (i.e. $I_b$) which directly controls the output.

Thus negative feedback occurs in an electronic circuit or system when some of the system's output signal is fed back to the input so as to act in opposition to its input signal. Negative feedback is a very common technique in the design of amplifier circuits, for it not only stabilises the operation of the amplifier (as in the circuit just decribed) but it is also used to reduce signal distortion and to reduce the effect of a few other problems as well. Negative feedback is discussed again in
△ Chapter 5, Book D.

▽ 25.2   **Motor speed controller**

A second example of how negative feedback is used to good effect in a transistor circuit is shown in figure 25.2. Here the emitter current passing through a d.c. motor is controlled by adjustment of the base current with the variable resistor, $VR_1$. This circuit is the basis for the Project Module A1, Motor Speed Controller, described in Book A.

$VR_1$ is wired as a potentiometer across the power supply so that the voltage on the base of $Tr_1$ can be adjusted. Any voltage, $V_{be}$, greater than about 0.7 V will switch on the transistor. But instead of switching the motor fully on, as would happen if the motor were connected in the collector circuit of the transistor, a smooth control of speed is obtained. Negative feedback is responsible for this behaviour.

As soon as $V_{be}$ reaches about 0.7 V and the transistor switches on, current flows through the motor. As the motor begins to turn, the voltage across it rises, largely due to back e.m.f., so that the voltage at point X rises, tending to switch off the transistor by decreasing the base-emitter voltage. But as this happens the motor speed reduces, decreasing the back e.m.f. and hence increasing $V_{be}$.

This is negative feedback in action: any increase in the output (here emitter current and hence motor speed) generates a signal (here the voltage on the emitter of $Tr_1$) which decreases the output current. So what happens is that the speed of the motor is held steady, not able to increase or decrease because of the controlling voltage on the emitter of the transistor. What is more, if the motor is called upon to provide more power it will slow down. In slowing down the back e.m.f. decreases so the transistor switches further on providing more current and more power. So the motor tends to run at a steady speed under varying loads.

$V_{be}$ changes by only a small amount of its input characteristic (see figure 24.3) while $I_e$ changes by a large amount as the base voltage is varied between about 0.7 V and the power supply voltage. The circuit is just as effective at controlling the brightness of a filament lamp since the resistance of the lamp increases as the current through it increases and provides
△ the negative feedback.

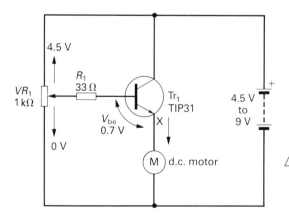

**Figure 25.2** Circuit for a motor speed controller which uses negative feedback

## ▽ 25.3 The transistor oscillator and positive feedback

The circuit shown in figure 25.3 is called an *astable multivibrator*. This is a type of electronic oscillator which continuously produces a rectangular, or square, waveform. The term 'multivibrator' refers to the fact that a rectangular waveform, taken from the collector of either transistor, consists of a very large number of different frequencies. It is the combination of these harmonics which produces the rectangular waveform. Astable multivibrators can also be built using integrated circuits, e.g. the 555 timer (see Chapter 14, Book B), and the 741 operational amplifier (see Section 3.6, Book D).

In the transistor version of the astable multivibrator, the two transistors are coupled together in such a way that while one transistor is conducting the other is non-conducting. The switch-over from one transistor conducting to the other conducting is entirely automatic and controlled by circuit action called *positive feedback*.

The charging time of each capacitor, and hence the frequency of the waveform, is determined by the time constant of the $R_1 \times C_1$ and $R_2 \times C_2$ combinations. The frequency, $f$, of the waveform produced at either collector is given by the formula

$$f = \frac{1}{(t_1 + t_2)}$$

$t_1$ is the time for which $Tr_1$ is on and $Tr_2$ is off per cycle time $(t_1 + t_2)$; $t_2$ is the time for which $Tr_1$ is off and $Tr_2$ is on per cycle time.

$$t_1 = 0.7 \ C_1 R_1 \text{ and } t_2 = 0.7 \ C_2 R_2$$

If $R_1 C_1 = R_2 C_2$, then

$$f = \frac{1}{1.4 \ R_1 C_1} \text{ or } \frac{1}{1.4 \ R_2 C_2}$$

Figure 25.4 shows a practical transistor astable which is designed to make two light emitting diodes, $LED_1$ and $LED_2$, flash on and off alternately. Each LED and its series resistor could be replaced by a 6 V, 60 mA lamp if a brighter flash is needed. In this circuit, $C_1 = C_2 = 100 \ \mu F$, and $R_1 = R_2 = 10 \ k\Omega$. Using the above equation, each LED flashes with a frequency

$$f = \frac{1}{(1.4 \times 100 \times 10^{-6} \times 10 \times 10^{3})}$$

$$= 0.6 \ Hz \text{ approximately.}$$

This is about twice per second. Remember when using electrolytic capacitors that they have such a wide tolerance that the values marked on them are only approximate. Therefore calculations can only be approximate. If the value of the capacitors is reduced, say by a factor of 1000, i.e. $C_1 = C_2 = 0.1 \ \mu F$, the frequency of oscillation is increased 1000 times to 600 Hz. Now at this frequency the LEDs will appear to be on continuously. However, the frequency of oscillation will now produce an audio tone in a

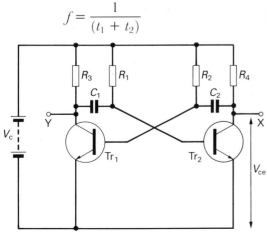

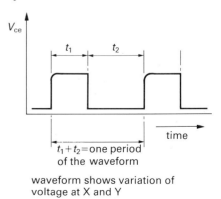

$t_1 + t_2$ = one period of the waveform

waveform shows variation of voltage at X and Y

**Figure 25.3** The two-transistor astable multivibrator

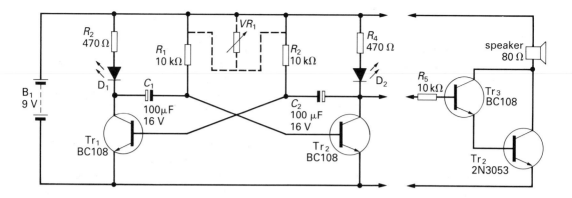

**Figure 25.4** A practical transistor astable

loudspeaker. A good way to drive a speaker with this astable is to connect a Darlington pair to the collector of either transistor as shown in figure 25.4.

The Darlington pair enables you to drive a relay, or turn a small d.c. motor on and off when the astable is operating at low frequency. The easiest way to control the frequency of the astable is to use a variable resistor, $VR_1$, as shown by the dotted lines. The transistor astable is a low-cost way of producing on/off signals for flashing lamps, controlling relays and producing audio tones. However, the integrated circuit '555 timer' (Book B, Chapter 14 and Book D, Chapter 9) is a rather easier △ way to produce rectangular waveforms.

## 25.4  How the two-transistor astable works

The working of the two-transistor astable can be understood from figure 25.5.

The sudden surge of current in the circuit when the power supply is first switched on, will quickly establish oscillations. Let us suppose that transistor $Tr_2$ is just switching on and $Tr_1$ is just switching off. This means that the voltage, $V_{ce}$, on the collector of $Tr_2$ is just about to fall to 0 V, and the voltage $V_{be}$ on the base of $Tr_1$ is just about to fall below 0.7 V. This condition of the circuit is shown by point X on the waveforms in figure 25.6.

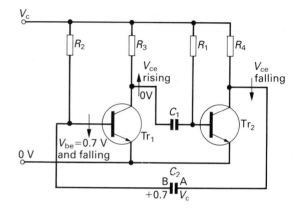

**Figure 25.5** Analysing the function of the transistor astable

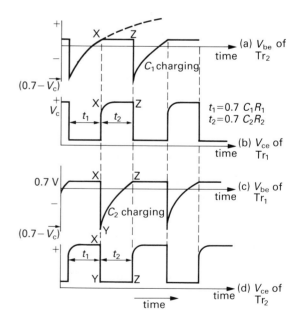

**Figure 25.6** Waveforms at the collectors and emitters of the astable multivibrator

Since capacitor $C_2$ is connected between the collector of $Tr_2$ and the base of $Tr_1$, the voltage across its terminals A and B is $(V_c - 0.7)$ V. Now when $Tr_2$ switches on, its collector voltage falls smartly to just above 0 V (point Y on figure 25.6(d)), and with it the voltage on terminal A of $C_2$ falls by an amount $V_c$. Since the capacitor has no time to adjust its charge, the voltage on terminal B, and hence the base-emitter voltage of $Tr_1$, must also fall by $V_c$. In other words the base-emitter voltage of $Tr_1$ falls to $-(V_c - 0.7)$ V (point Y on figure 25.6(c)). Thus the circuit actually generates a negative voltage, i.e. a voltage below 0 V!

This negative voltage positively switches off $Tr_1$ and its collector voltage sharply rises to $V_c$ (point X on figure 25.6(b)). This action is known as *positive feedback* since the falling voltage on the collector of $Tr_2$ is conveyed by capacitor $C_2$ to produce a falling voltage on the base of $Tr_1$. We can regard the collector-emitter voltage of $Tr_2$ as the output voltage, and the base-emitter voltage of $Tr_1$ as the input voltage.

constant output voltage, in a circuit; positive feedback causes instability, i.e. oscillations, of the output voltage.

Of course, the negative voltage on the base of $Tr_1$ cannot remain at $-(V_c - 0.7)$ V since $C_2$ must charge up through $R_2$. The rising voltage on terminal B of $C_2$ is shown by the curve from Y to Z in figure 25.6(c). The voltage on terminal B would normally reach $V_c$ but only reaches about 0.7 V for this is when $Tr_1$ switches on. At this instant, the collector voltage, $V_{ce}$, of $Tr_1$ suddenly falls (figure 25.6(b)), while that of $Tr_2$ suddenly rises (figure 25.6(d)). Positive feedback now takes place between the collector of $Tr_1$ and the base of $Tr_2$ via positive feedback acting through capacitor $C_1$; $Tr_2$ then smartly switches off. Now $C_1$ charges up through $R_1$ and so the cycle continues.

If $C_1 \times R_1 = C_2 \times R_2$ the times $t_1$ and $t_2$ are equal. For the waveform at either collector, the mark (on time) to space (off time) ratio is equal, i.e. $t_1/t_2 = 1$. Clearly, by varying the $CR$ values, different mark-to-space (M/S) ratios can be obtained.

$\triangle$

## Remember

Positive feedback occurs if part or all of the falling (or rising) voltage at the output is fed back to the input to make the input signal assist the falling (or rising) output voltage. We say that the input and output signals are *in phase*.

You should compare this action with the effect of negative feedback which was discussed in Section 25.1 in relation to stabilising the operation of a transistor amplifier.

Negative feedback occurs if part or all of the falling (or rising) voltage at the output is fed back to the input to oppose the falling (or rising) output signal. We say that the input and output signals are *out of phase*.

Negative feedback causes stability, e.g.

25.5  **The Schmitt trigger**

The transistorised switching circuits described so far in this book don't perform quite as well as they might: they lack 'snap-action'. Even the Darlington pair (Chapter 20) hasn't got this desirable ingredient, though it performs its function as a sensitive electronic switch well enough.

'Snap-action' describes a circuit which switches on at one particular voltage and switches off at a slightly different voltage. A mechanical analogy of this behaviour is a wall light switch: press it down, and it suddenly springs on; press it up and it suddenly springs off. Just like the wall switch, the action of the Schmitt trigger relies on positive feedback. The basic arrangement of a Schmitt trigger is shown in figure 25.7.

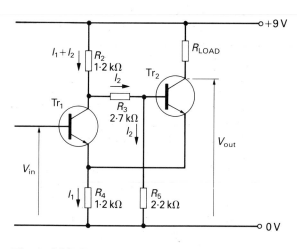

**Figure 25.7** The essence of a Schmitt trigger

Two voltages are important in the operation of the Schmitt trigger: one is the voltage $V_{on}$ applied to the base of transistor $Tr_1$ which causes this transistor to switch on; the second is the lower voltage $V_{off}$ which causes this transistor to switch off. The difference between these voltages is known as *hysteresis*. The word means 'lagging behind' and in this case means that $V_{off}$ lags behind $V_{on}$. Hysteresis makes switching circuits considerably more useful in control systems, e.g. the thermostat described in Section 21.2 — they provide 'snap-action'. From figure 25.7 we can calculate the two switching voltages which are responsible for 'snap-action'. Firstly, $V_{on}$:

Suppose $Tr_1$ is off and $Tr_2$ is on. Assume that the base current of $Tr_2$ is small compared with the current flowing through the voltage divider chain made up of $R_2$, $R_3$ and $R_5$. You should be able to see that the base voltage of $Tr_2$ is given by:

$$9 \ R_5/(R_2 + R_3 + R_5)$$

assuming that the supply voltage is 9 V.

If we neglect the small base-emitter

voltage of $Tr_2$ (about 0.7 V for a silicon transistor), this voltage is the voltage at the emitter of $Tr_2$ and hence at the emitter of $Tr_1$, since these two transistors are emitter-coupled. Again neglecting the small voltage across the base-emitter of $Tr_1$ when this transistor stitches on, the equation also gives the voltage required to switch on $Tr_1$. For the values of the three resistors in figure 25.7, $V_{on}$ is therefore close to 3.3 V. If the input voltage is increased to a small value just above 3.3 V, positive feedback will drive $Tr_1$ more strongly on and $Tr_2$ more strongly off.

Now for the calculation of $V_{off}$. It is not likely that $V_{off}$ will be equal to $V_{on}$ since quite a violent change in the circuit has caused $Tr_1$ to switch on and $Tr_2$ to switch off. However, suppose you have reduced the base-emitter voltage of $Tr_1$ so that $Tr_2$ is just about to switch on. Since $Tr_1$ is already on and $Tr_2$ is just about to come on, the voltages at the bases of both transistors are close to their respective emitter voltages, since they are connected together. Thus the emitter and base terminals of these transistors are close to the voltage $V_{off}$ we want to calculate. Then:

$$I_1 = V_{off}/R_4 \text{ and } I_2 = V_{off}/R_5$$

and the voltage drop across $R_2$ and $R_3$ is given by:

$$9 - V_{off} = (I_1 + I_2) \ R_2 + I_2 R_3$$

Now if you substitute the values of $I_1$ and $I_2$ given above, you will obtain the equation for $V_{off}$:

$$V_{off} = 9/(1 + (R_2/R_5) + (R_2/R_4) + (R_3/R_5))$$

Substitution of the resistor values in figure 25.7 gives a value of $V_{off} = 2.4$ V. Thus the hysteresis of the circuit is $V_{on} - V_{off} = 3.3$ V $- 2.4$ V $= 0.9$ V. Clearly different values of $R_2$ to $R_5$ provide different values for the hysteresis.

So what are the practical advantages of using a Schmitt trigger instead of a Darlington pair? Look at the graphs shown in figure 25.8. The top one shows an irregular series of input pulses applied to

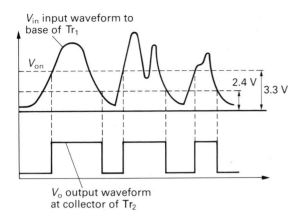

**Figure 25.8** The action of a Schmitt trigger

the base of Tr$_1$. The Schmitt trigger switches on when the amplitude of the input pulses exceeds 3.3 V, and switches off when it falls below 2.4 V. The output pulses are of constant amplitude and switch on and off sharply. Now this is a useful function.

One application is the use of a Schmitt trigger as a discriminator; that is, it will discriminate between pulses of different amplitudes — all pulses below a certain amplitude will not 'fire' the Schmitt trigger. The discrimination can easily be set by adjusting the voltage divider across the base of Tr$_1$. Thus the Schmitt trigger is useful in pulse-height analysis circuits for counting, say, the distribution of energies of cosmic rays. But a more 'earthly' use is in control circuits, an example of which is the thermostat described in Section 21.2. The transistorised Schmitt trigger shown in figure 25.7 is the basis of Schmitt triggers in integrated circuit packages which are described in Book D, Chapter 4 and Book E, Chapter 9.

# 26 Field-effect Transistors

## 26.1 Introduction

The properties and uses of only two types of bipolar transistors, the npn and pnp types, have so far been discussed. The field-effect transistor (FET) is not so widely used as the bipolar transistor, but it does form the basic building block of CMOS integrated circuits which are discussed in Books D and E. (CMOS stands for *complementary metal-oxide semiconductor*.)

Field-effect transistors, in either discrete or integrated form, are preferred for three main reasons: their lower power requirements, their higher input impedance, and their higher frequency-response compared with bipolar transistors. Their lower power requirement makes field-effect transistors a good choice for integrated circuits since excessive heat is not produced when many thousands of FETs are integrated on a small area of a silicon chip. It is therefore possible to design portable battery-operated devices, e.g. personal computers and solar-powered calculators using this type of integrated circuit.

One popular type of field-effect transistor is the VMOS (vertical metal-oxide semiconductor) type. It is used in Project Module A5, the Audio Amplifier and Project Module B2, the Relay Driver. Its properties and uses form the main part of this chapter.

Interestingly, the FET was invented in the 1930s, a decade before the invention of the bipolar transistor, as result of a search for a solid-state equivalent of the triode valve. The invention was not taken up commercially, largely because at the time it was not possible to produce semiconductor materials of sufficient purity.

## 26.2 What FETs look like

As figure 26.1 (overleaf) shows, field-effect transistors look just like bipolar transistors. They have three terminals, the *drain* (d), *source* (s) and *gate* (g). There are two main types of FET: the JUGFET (junction-gate field-effect transistor) and the MOSFET (metal-oxide silicon field-effect transistor). Figure 26.1 shows an example of each of these types of FET, and their circuit symbols. The 2N3819 is a JUGFET and the VN46AF is a type of MOSFET, actually a VMOS type. Since the relatively new VMOS FET is now available at a reasonable cost, and it has some interesting uses, we will not be greatly concerned with the JUGFET.

The following two simple experiments show the potential of the field-effect transistor, in this case, the VMOS type of MOSFET.

Just as there are npn and pnp bipolar transistors in which current flow is due to both electrons and holes, there are also n-channel and p-channel FETs in which the current flow is due to other electrons or holes, respectively. N-channel and p-channel FETs, especially VMOS types of MOSFETs, are used as complementary-pairs in audio amplifiers. But only n-channel FETs are discussed in this section. The simple difference between the circuit symbols for the p-channel and n-channel FETs is that the arrows are pointing the opposite way — see figure 26.1.

**Figure 26.1** Details of three types of field-effect transistor

| field-effect transistor | symbol | pin identity (pin view) | general appearance |
|---|---|---|---|
| 2N3819 n-channel junction gate field-effect transistor (JUGFET) | 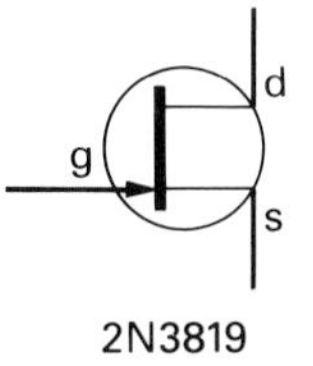 2N3819 | 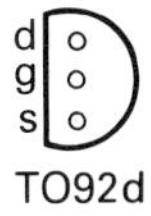 d g s TO92d | 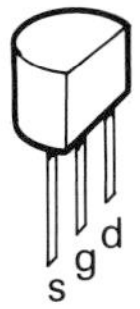 d g s |
| VN10LM n-channel metal-oxide semiconductor field-effect transistor (MOSFET) (VMOS type) | 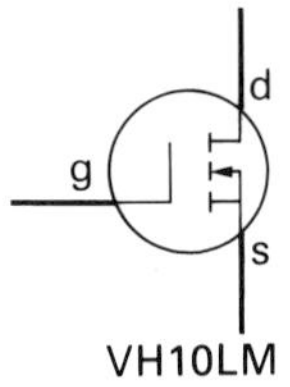 VH10LM | 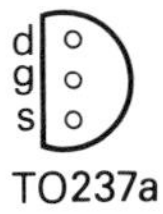 d g s TO237a | 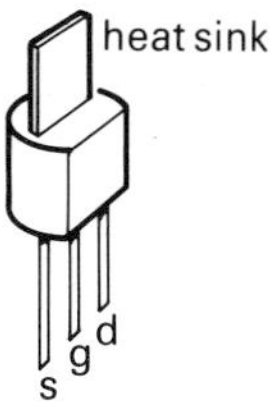 heat sink d g s |
| VN46AF VN66AF (MOSFET) (VMOS type) | 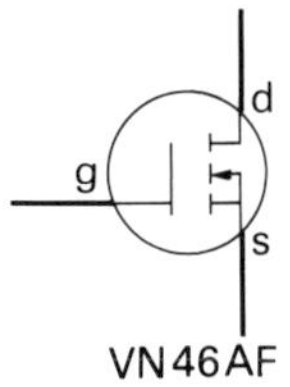 VN46AF | 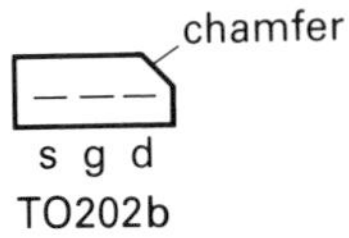 chamfer s g d TO202b | 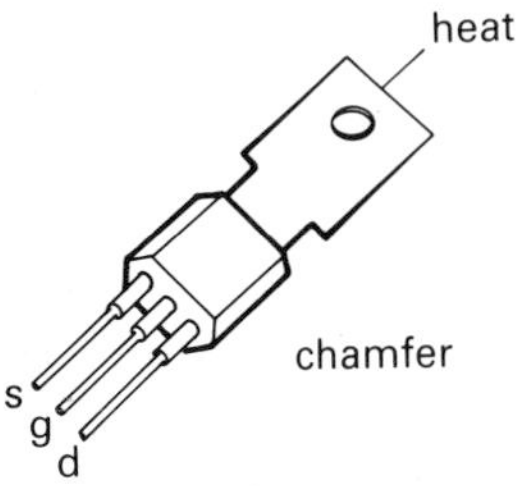 heat s g d chamfer |

## 26.3 *Experiment* C14

### Making an FET touch switch

Figure 26.2 shows how a VMOS FET can be used as a simple touch switch. Use a VN10LM, a VN46AF, or a VN66AF for $Tr_1$. Once the circuit is assembled on breadboard or terminal block, the lamp can be switched on by bridging the finger across the two bare contacts, A and B.

Before the contacts are touched the lamp is off, and resistor $R_1$ holds the gate terminal of $Tr_1$ at 0 V. Clearly by bridging the contacts you raise the voltage at the gate of the FET, and this allows current to flow through the transistor to light the lamp. How the FET does this is explained in Section 26.6.

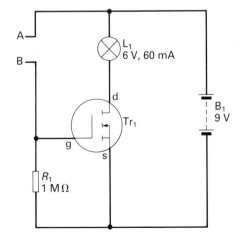

**Figure 26.2** Experiment C14: Touch switch

# Questions

1 How could you use this circuit to energise a relay when the contacts are bridged?
2 Remove the 1 MΩ resistor. Comb your hair then move the comb near the circuit. You should see the brightness of the lamp vary. This effect is caused by the voltage on the electrostatically-charged comb. This voltage creates an electric field which alters the current flowing through the field-effect transistor.

## 26.4 *Experiment* C15

## Making an FET timer

Figure 26.3 shows how a VMOS FET can be used as a simple timer. Once the circuit is assembled on a breadboard or terminal block, remove the link wire between A and B and the lamp will light after a time delay. The time delay, $T$, is very approximately given by the equation

$$T = 0.3 \times C_1 R_1 \text{ seconds}$$

if $C_1$ is in farads and $R_1$ in ohms. Thus for the values in figure 26.3 the time delay ought to be about

$$0.3 \times 500 \times 10^{-6} \times 100 \times 10^3 = 15 \text{ s}$$

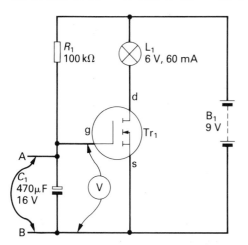

**Figure 26.3** Experiment C15: Timer

# Questions

1 How could you use this circuit to energise a relay after the link wire is removed?
2 How would you modify the circuit so that the lamp goes out after a time delay?
3 Connect a 10 V d.c. voltmeter, V, as shown and record the variation of voltage across the capacitor, i.e. across the gate and source terminals of $VR_1$. What happens to the reading when the lamp comes on? Is this behaviour different from what you would get if $Tr_1$ were replaced by a bipolar transistor?

The reason why the touch switch and timer circuits work as they do depends on the special way FETs are made. First let's look at the structure and operation of the JUGFET.

▽ 26.5   **The structure of the JUGFET (junction-gate field-effect transistor)**

Figure 26.4 (overleaf) shows the essential structure of a JUGFET. It comprises a channel of n-type semiconductor through which a current (of electrons) flows between the source of electrons (terminal s) and the 'drain' for electrons (terminal d). This current, $I_{ds}$, is driven along by the drain-to-source voltage, $V_{ds}$. The size of $I_{ds}$ is controlled by the gate-to-source voltage, $V_{gs}$. The gate terminal, g, is connected to p-type semiconductor material which is embedded in the n-type channel. Note that the gate terminal is more negative than the source terminal. Note also, that although electrons flow from source to drain, the conventional current, $I_{ds}$, flows from drain
△ to source.

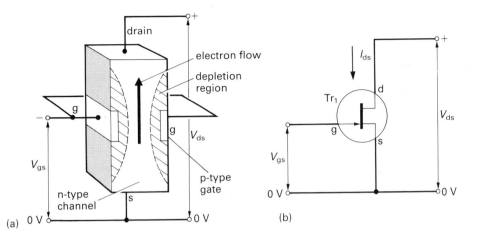

**Figure 26.4** (a) Structure and (b) voltages controlling current through an n-channel JUGFET

▽ 26.6 **How JUGFETs work**

Since the gate terminal of a JUGFET is more negative than its source terminal, the p-n junction between the p-type gate and n-type channel is reverse-biased. This reverse-biased p-n junction forms a depletion region extending into the n-type channel as shown in figure 26.5. Since the depletion region is depleted of electrons in the n-type channel, it has a high resistance, and current is restricted to the remaining conducting n-type channel. The width of this conducting channel is dependent on the size of the gate-source voltage, $V_{gs}$.

Figure 26.5 compares the width of the n-type channel, and hence its resistance, for two different values of $V_{gs}$, i.e. $V_{gs} = -1$ V and $-2$ V. The higher $V_{gs}$, the narrower the channel; the higher its resistance, the lower $I_{ds}$. Thus, in principle, the action of the JUGFET is simple: it contains a conducting channel made of n-type or p-type material. The resistance of this channel, and hence the size of the current flowing through it, is controlled by voltage applied to the gate which varies the width of the channel.

Note the pear-shaped structure of the depletion region in figure 26.5. It shows that the depletion region extends deeper

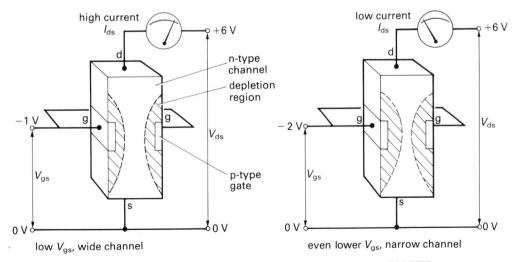

**Figure 26.5** How the depletion region controls the current through an n-channel JUGFET

into the n-type channel towards the drain terminal. This is because the gate terminal becomes increasingly more negative with respect to the channel, in moving from the source to the drain.

The principal difference between the JUGFET and the bipolar transistor is the formation of this reverse-biased p-n junction in the n-type channel. Whereas the bipolar transistor needs a small base current across the forward-biased base-emitter junction to make it work, the unipolar JUGFET has (but does not need) a very small gate current across the reverse-biased gate-source junction. In other words, the JUGFET (and all other types of FET) has a much higher input resistance than a bipolar transistor.

The high input resistance of the FET is shown by experiments C14 and C15. When you moved the charged comb near the touch switch with the 1 MΩ resistor removed, the electric field surrounding the comb was able to influence the gate-source voltage even though the electric field provided negligible gate-source current. You wouldn't get the same effect with a bipolar transistor.

When you measured the voltage, $V$, across the gate-source terminals in the timer circuit, the voltage continued to rise after the lamp lit. In a timer circuit based on a bipolar transistor (e.g. figure 26.5), a base current has to flow in order to switch on a collector current. However, in the timer based on the FET, there is negligible gate current when the drain current is high enough to light the lamp. Consequently, the gate-source voltage goes on rising after the FET has switched. The base-emitter voltage of a silicon bipolar transistor would rise to about 0.6 V and stay at this value as current flows through the base-emitter junction.

Many n-channel JUGFETs have the structure shown in figure 26.6. Individual transistors are built up on a silicon substrate by the epitaxial process. This involves the growth of thin layers of n-type and p-type silicon on a silicon substrate by heating the substrate in different gases,

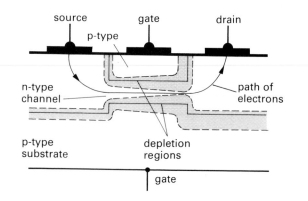

**Figure 26.6** The structure of an n-channel FET using the epitaxial planar construction

e.g. phosphene gas to give n-type silicon. The gases diffuse into the surface of the silicon to give a U-shaped n-channel FET. The electric field produced by the gate-source voltage produces a depletion region in the channel. The epitaxial process is used to make the many thousands of transistors on a silicon chip.

▽ 26.7 # The structure of the MOSFET (metal-oxide semiconductor field-effect transistor)

MOSFETs are made by a variation of the epitaxial process shown in figure 26.6 to give a much higher input resistance than the JUGFET. The gate of a MOSFET is electrically insulated from the channel by a very thin layer of silicon dioxide. The electric field generated by the gate-source voltage is still able to influence the resistance of the channel, but in this case barely any current at all passes through the gate terminal. The input resistance of a MOSFET is extraordinarily high, so high, in fact, that the resistance between the copper tracks on a printed circuit board is likely to be lower than the input resistance of the MOSFET!

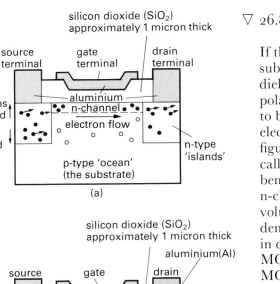

**Figure 26.7** The structure of two types of n-channel MOSFET   (a) n-channel enhancement MOSFET in which the conducting n-channel is induced during operation   (b) n-channel depletion MOSFET in which the conducting n-channel is produced during manufacture

▽ 26.8 **How MOSFETs work**

If the gate is positive with respect to the substrate, the silicon dioxide acts as the dielectric in a parallel-plate capacitor. The polarisation of the oxide layer causes holes to be repelled from the oxide layer and electrons to be attracted to it as shown in figure 26.7(a). Thus a layer of electrons, called an *inversion layer*, is formed just beneath the oxide layer which acts as the n-channel. Changes to the gate-to-source voltage, $V_{gs}$ cause changes to the electron density in this channel and hence changes in drain to source current $I_{ds}$. This type of MOSFET is called an enhancement MOSFET since an electron-rich n-channel must first be produced by applying a positive voltage to the gate. The MOSFETs shown in figure 26.1 are enhancement MOSFETs. Indeed, enhancement MOSFETs are used extensively in digital and analogue circuits. Since the channel doesn't exist until a positive gate voltage is applied, this MOSFET is also called a *normally-off MOSFET*. These n-channel MOSFETs are also called *n-MOS transistors*. Note that n-channel MOSFETs differ from n-channel JUGFETs in that the latter requires a gate voltage which is negative with respect to the source. The n-channel enhancement MOSFET produces drain-to-source current which is enhanced by the gate-to-source voltage.

The depletion mode MOSFET shown in figure 26.7(b) is usually made as a discrete component. Unlike the enhancement MOSFET, it already has a conducting channel when the gate-to-source voltage, $V_{gs}$, is zero since a thin n-channel is diffused in between the gate during manufacture. Thus for the n-channel device shown in figure 26.7(b), a negative $V_{gs}$ reduces the channel width so that $I_{ds}$ decreases (just like the n-channel JUGFET). By making $V_{gs}$ positive, the channel width can be increased, so increasing $I_{ds}$. This type of MOSFET is known as a *normally-on MOSFET*.

For p-type MOSFETs, both

Figure 26.7 shows two types of n-channel MOSFET, the *enhancement MOSFET* and the *depletion MOSFET*. Both types are made starting with an 'ocean' of p-type substrate in which two heavily-doped 'islands' of n-type material are diffused to provide the source and drain regions. These islands are actually very close together — typically about 0.005 mm. A layer of aluminium is deposited over the source and drain regions to which external connections are made. Next a layer of silicon dioxide is formed over the small gap between the source and drain. Aluminium is then deposited over the silicon dioxide layer to △ make the gate connection.

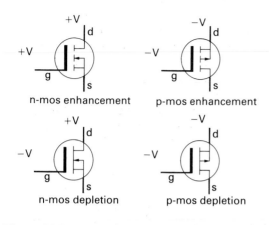

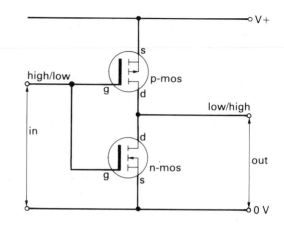

**Figure 26.8**  The symbols for the four types of MOSFET

**Figure 26.9**  The structure of a CMOS

enhancement and depletion types, the dopings shown in figure 26.7 are reversed. The symbols for the four kinds of MOSFET are shown in figure 26.8. The arrow in each symbol indicates direction of current flow in the forward-biased p-n junction formed between the substrate and the source on drain regions. Note the distinction between the normally-off and normally-on types: a continuous line represents the channel for normally-on, △ and a broken line for normally-off.

▽ 26.9  **The structure of CMOS (complementary metal-oxide semiconductor)**

CMOS refers to the development of integrated circuits which combine both p-channel and n-channel enhancement MOSFETs on the same silicon chip.

An example of an 'integrated circuit' which can be made using a complementary pair of MOSFETs is the logic inverter (a NOT gate) shown in figure 26.9. Here the p-channel and n-channel transistors are in series. The output signal is taken from the common connection of the two drain

terminals, and the input signal is applied to both gates simultaneously. This NOT gate works as follows.

Suppose a high input signal (e.g. having a d.c. voltage of 10 V) is applied to the input. This voltage turns on the n-channel transistor and turns off the p-channel transistor. Thus the resistance of the drain of the n-channel transistor will be very low and the resistance of the p-channel will be very high so that the output voltage is low (i.e. close to 0 V). On the other hand, a low input signal (i.e. having a d.c. voltage of 0 V) switches on the p-channel transistor and turns off the n-channel transistor so that the output voltage goes high (i.e. near the supply voltage). This is the action of a CMOS inverter or NOT gate.

As explained in Book E, CMOS digital integrated circuits have a low power consumption, and operate from a supply voltage in the range 3 V to 15 V (and some to 18 V). They have a very low input current (e.g. 10 pA), and they can be packed close together on a silicon chip. It is for some of these reasons that CMOS ICs are used in many of the practical circuits throughout *Basic Electronics*, including many of the Project Modules — △ see Book A, Chapter 12.

## ▽ 26.10  The structure and operation of a VMOS transistor

The structure of a VMOS (vertical metal-oxide semiconductor) FET is unlike that of a conventional MOSFET transistor which has the drain, source and gate connections on the top of the silicon chip. Figure 26.10 shows the structure of an n-channel VMOS transistor. Note the 'vertical' construction. The VMOS transistor has its source on the top of the chip and its drain at the bottom. The drain terminal is connected to a heavily doped (n$^+$) substrate which allows heat to be removed efficiently via a heatsink connected to the substrate. A lightly doped p$^-$ region, called the body, and an n$^+$ region for the source are diffused into this epitaxial layer.

A feature of the VMOS FET is the V-groove which is etched through the p$^-$ and n$^+$ regions into the epitaxial layer. Aluminium is deposited on an insulating layer of silicon dioxide which is grown over the groove. Similar connections are made for the source and drain by depositing aluminium over the p$^-$ region

for the source and the n$^+$ region for the drain. VMOS is an enhancement MOSFET so current flows through it vertically from the drain to source along both sides of the V-groove gate, but only when both the drain and gate are positive with respect to the source.

The main advantage of the VMOS FET compared with the planar construction of a conventional MOSFET is its larger current, voltage and power rating. Since the vertical process produces channels, each about 1.5 mm long, its current capacity is doubled. Also, because the substrate forms the drain contact, heat can be removed efficiently, and its breakdown voltage is higher than conventional MOSFETs. Additionally, VMOS FETs, like all MOSFETs, have a very high input resistance. Their input currents of less than 100 nA mean that VMOS can be directly interfaced with CMOS logic — see Section 28.6.

Also the VMOS FET is a unipolar device, i.e. conduction through it is by majority carriers, electrons or holes. The control of these carriers is by a field, not by the injection or extraction of minority carriers as with bipolar transistors. Thus VMOS FETs switch current on and off faster than bipolar transistors which are slower because of the recombination time of electrons and holes. Some VMOS devices can switch a 1 A current on and off in less than 4 ns.

Finally, VMOS devices have a negative temperature coefficient while bipolar transistors have a positive temperature coefficient. Thus where bipolar transistors can suffer damage by thermal runaway, VMOS draws less currrent as the device heats up: if the temperature increases at a particular point in the device, the current decreases and the temperature falls. Thus current automatically equalises throughout the chip, i.e. there are no 'hot spots'. Similarly, when VMOS devices are connected in parallel to increase current switching capacity, the current is △ automatically shared between the devices.

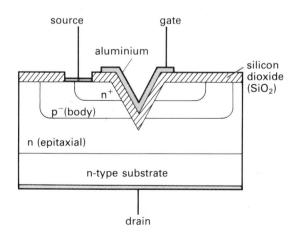

**Figure 26.10** The structure of an n-channel enhancement VMOS transistor

## ▽ 26.11   **FET characteristics**

Typical output characteristics for an n-channel VMOS transistor are shown in figure 26.11. The curves for an n-channel JUGFET are similar except that $I_{ds}$ is smaller, and $V_{gs}$ is negative. Note the following:

**(a)** For each value of $V_{gs}$, $I_{ds}$ increases linearly for small increases in $V_{ds}$. This is the region where the channel acts as an Ohm's law resistor — its resistance depending on the value of $V_{gs}$. The resistance of the drain of a VMOS transistor varies from a few ohms for high values of $V_{gs}$ to several hundred ohms (low values of $V_{gs}$).

**(b)** The 'knee' of each curve indicates the point where a further increase in $V_{ds}$ saturates $I_{ds}$, i.e. the channel has become so narrow that any increase in $I_{ds}$ (expected with a continued increase in $V_{ds}$) is balanced by the reduction due to channel narrowing. Above the knee the VMOS is said to be in the pinch-off region. $V_p$ is the pinch-off voltage defined as the value of $V_{ds}$ for which $I_{ds}$ is a maximum. Maximum permitted values for $I_{ds}$ are shown in figure 26.12

**(c)** If $V_{ds}$ is increased beyond a certain value, breakdown occurs in the channel, $I_{ds}$ increases sharply, and the VMOS will probably be damaged. Figure 26.12 shows maximum permitted values of the drain-source voltage of some VMOS transistors.

Figure 26.13 shows how the drain current, $I_{ds}$, varies with gate-source voltage, $V_{gs}$ for a given drain-source voltage $V_{ds}$. This curve is typical of the input, or transfer, characteristic of a VMOS device.

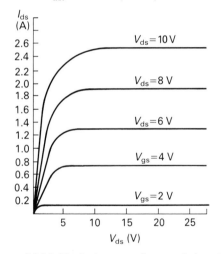

**Figure 26.11**  Typical output characteristics of a VMOS device

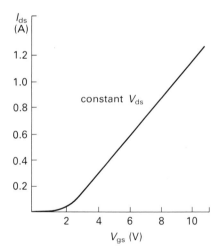

**Figure 26.13**  Typical input characteristic of a VMOS device

| device | max $I_{ds}$ (A) | max $V_{ds}$ (V) | minimum $R_{ds}$ (Ω) | maximum power (W) | minimum $g_m$ (mS) |
|---|---|---|---|---|---|
| VN10CM | 0.5 | 60 | 5 | 1 | 100 |
| VN46AF | 2 | 40 | 3 | 15 | 150 |
| VN66AF | 2 | 60 | 3 | 15 | 150 |
| VN88AF | 2A | 80 | 4 | 15 | 150 |

**Figure 26.12**  Ratings of a few n-channel VMOS transistors

Note the following:

(a)  The gate-source voltage required to switch on the VMOS transistor is higher than for a bipolar silicon transistor, and is usually between about 0.8 V and 2 V.

(b)  The slope of this curve, i.e. $I_{ds}/V_{gs}$, is known as the transconductance, $g_m$, of the VMOS (and all other FETs, too).

Transconductance is usually expressed in siemens, S, since the reciprocal of resistance is conductance, measured in siemens. In the examples in figure 26.12 transconductance is expressed in millisiemens (mS). A knowledge of transconductance is useful in working out the voltage gain of audio amplifiers using

$\triangle$ FETs.

TECHNOLOGY WORKSHOP
CORNWALL COLLEGE
POOL, REDRUTH,
CORNWALL TR15 3RD

# 27 Thyristors and Triacs

27.1 **Introduction**

These devices are used in power control circuits and, like rectifiers, depend on the properties of the p–n junction. They are to be found in foodmixers, electric drills and lamp dimmers. Figure 27.1a shows the appearance of a typical *thyristor*. This device was formerly known as a *silicon controlled rectifier (s.c.r.)* since it is a rectifier that controls the power delivered to a load, e.g. a lamp or motor. The symbol for a thyristor is shown in figure 27.1b: it looks like a diode symbol with anode and cathode terminals, but with the addition of a third terminal called a gate, g. A simple demonstration of d.c. power control using the thyristor, Thy, is shown in figure 27.1c.

This circuit forward-biases the thyristor but the thyristor does not conduct until a positive voltage is applied to the gate terminal by closing switch $SW_1$. This positive voltage allows a small current to flow into the gate terminal and the thyristor 'fires'. However, conduction continues and the lamp remains lit even though the positive gate voltage is removed by opening $SW_1$. The only way to switch the thyristor off is to open switch $SW_2$. Four layers of semiconductors make up the thyristor's p-n-p-n sandwich construction as shown in figure 27.1d. Note that the word thyristor is derived from the Greek thyra, meaning door, and indicates that the thyristor is either open or closed.

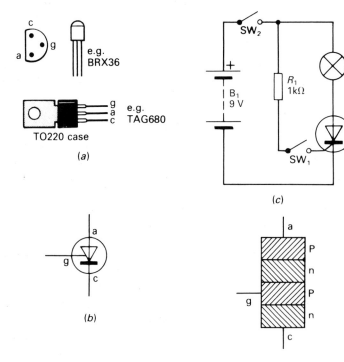

**Figure 27.1** The thyristor: (a) two types of package  (b) circuit sumbol  (c) a simple power control circuit  (d) internal construction

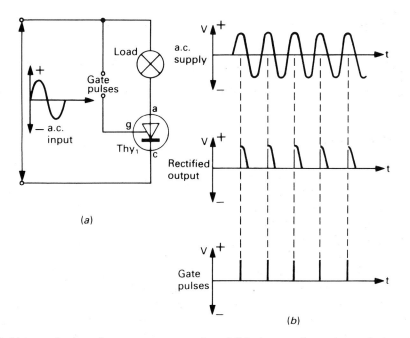

**Figure 27.2** (a) Using a thyristor for a.c. power control, and (b) the waveforms that make it work

## 27.2 Circuit applications

It is possible to use a thyristor to control a.c. power by allowing current to be supplied to the load during only part of each cycle. Figure 27.2 shows the basic circuit and waveforms. In a practical circuit, the gate pulses are applied automatically at a selected stage during part of each cycle. Thus half power to the load is achieved by applying the gate pulses at the peaks of the a.c. waveform. More or less power in the load is achieved by changing the timing of the gate pulses.

Since the thyristor switches off during the negative half cycles, it is only a half-wave device (like a rectifier) and allows control of only half the power available in a.c. circuits. A better device is a *triac*. This comprises two thyristors connected in parallel but in opposition and controlled by the same gate, i.e. it is bi-directional and allows current to flow through it in either direction. Figure 27.3 shows its symbol. The terms anode and cathode have no meaning for a triac; instead the contact near to the gate is called 'main terminal 1' ($MT_1$), and the other 'main

terminal 2' ($MT_2$). The gate trigger voltage is always referred to $MT_1$, just as it is referred to cathode in the thyristor. As with the thyristor, a small gate current switches on a very much larger current between the main terminals. A typical gate current is of the order of 20 mA but it is adequate for triggering triacs of up to 25 A rating.

If the triac is to be used in a lamp dimmer unit or a motor speed controller, there has to be some means of varying the a.c. power passing through the load. Figure 27.3b shows how this is achieved using an 'RC phase shifter'. $VR_1$ is the dimmer control. The device shown as a *diac* is in effect two zener diodes connected back to back. It conducts in either direction when the voltage across $C_1$ reaches the diac's breakdown voltage of about 3 V. The burst of current through the diac 'fires' the triac. The rate at which $C_1$ charges depends on the value of $C_1$ and that of the series resistor, $VR_1 (+ R_1)$. The greater the value of $VR_1$, the more slowly the capacitor charges and the later in each half-cycle is the lamp turned on and the dimmer the lamp.

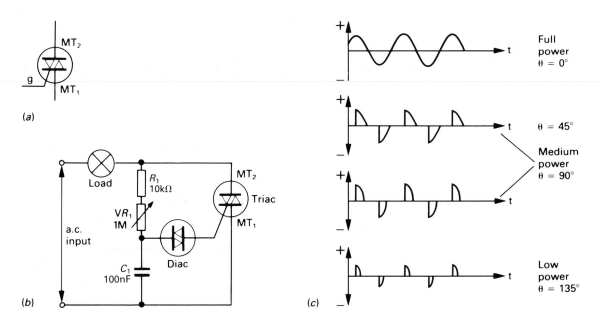

**Figure 27.3** The triac: (a) circuit symbol   (b) a basic circuit for a.c. power control   (c) the waveforms that make it work

The waveforms in figure 27.3c show how the triac controls power by chopping off part of each half cycle. The amount chopped off is indicated by the phase shift, $\theta$. If $\theta$ is 0, the triac conducts throughout the whole of each half-cycle of the a.c. waveform and the load is at full power. As $\theta$ increases from 0 to 180°, more and more of each half cycle is chopped off, and the dimmer the lamp becomes. Practical a.c. power control circuits using triacs refine the basic design in figure 27.3b to give better low-level power control, and a reduction of the radio interference that is generated by the rapid turn-on and turn-off of the triac. You might be interested to know that a quadrac combines the triac and diac in a single package.

# 28 Further uses for Transistors

## 28.1 Introduction

This section includes a number of applications for the bipolar and FET transistors described in this book. A brief explanation of how each circuit functions is included. You can build the circuits using either breadboard or stripboard. The leads of the transistors used in the circuits can be identified from figures 17.1 and 26.1.

## 28.2 Metal Detector

This simple application locates buried metal objects such as coins, buckles, flasks, and other metallic bric-a-brac of bygone ages. But archaeology apart, there are often urgent everyday reasons why a metal detector is useful — to find lost keys and other personal property, for instance. And supposing you want to find the position of nails in second-hand timber before sawing it; or to locate the path of electrical wiring or conduit before drilling a wall; or to trace the path of water pipes under the ground; or to decide whether your pet has swallowed the ring you have lost?

The circuit shown in figure 28.1 is very simple since part of the detector is an AM radio. The circuit is a radio transmitter known as a Colpitts oscillator. It produces a steady carrier wave, the frequency of which is near to the frequencies corresponding to medium waves on the AM band. This carrier wave interferes (or beats) with the carrier wave transmitted by a station on the medium wave band to produce an audible note.

In common with most metal detector circuits the frequency of the carrier wave produced by the circuit is affected by the presence of metal near the coil. This causes a change in the inductance of the coil, a corresponding change in the radio frequency carrier wave it produces, which, in turn, changes the audible note. What is more, the note changes one way for ferrous objects such as iron or nickel and the other way for non-ferrous objects such as brass, silver and aluminium.

The circuit makes use of positive feedback which is provided by the connection between the tapped capacitive divider ($C_1/C_2$) in the collector circuit, and the emitter connection of the transistor. Radio frequency oscillations are generated in the tuned circuit comprising $C_1$ and $C_2$, and the coil, $L_1$, and they are sustained by the amplifying action of the transistor, $Tr_1$.

This coil comprises about 100 turns of 28 s.w.g. enamelled copper wire wound on plastic or other non-metallic former about 150 mm in diameter. A handle is attached to this former and the circuit can be housed in a small box fixed to the handle so as to produce a compact device. A 9 V PP3 battery is suitable for operating the circuit. An insulated wire connected to the

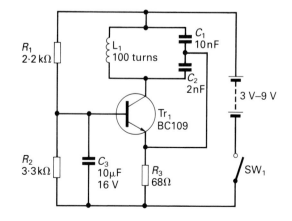

**Figure 28.1** Metal detector

positive line of the circuit is loosely wound round a portable radio.

Project Module B7 in Book B describes the practical construction of the Metal Detector.

Switch on the radio and tune in stations on the medium- or long-wave band. You should hear a whistle as you tune across a station, but choose a station where the whistle is loudest and the interference from speech or music is minimal. The note should not change in pitch as you tune across a station. Test the Metal Detector by bringing up various metal objects and experiment so that you can tell the difference between ferrous and non-ferrous objects.

**CAUTION**

Please note that at present there is considerable concern among professional archaeologists about the use of metal detectors by people who do not appreciate the archaeological value of their finds. If you are fortunate enough to discover metal objects, or any other objects which you suspect are of archaeological value, please contact your local museum about your find. In this way the value of the discovery can be gauged before a skilled dig is made.

## 28.3   Touch Volume Control

A touch volume control circuit illustrates the fact that the resistance of the channel of a field-effect transistor varies with the reverse-bias gate-source voltage. This controllable channel resistance is used to vary the voltage gain of an audio frequency preamplifier based on the bipolar pnp transistor, $Tr_1$. The voltage gain of this transistor is determined by the amount of a.c. decoupling provided by $C_4$ and the resistance of the channel of the $Tr_2$. If this impedance is low, the gain of the amplifier is high.

The drain-source voltage, $V_{gs}$, is determined by the state of charge of

capacitor $C_3$. Suppose $C_3$ is initially uncharged and contacts B and C are bridged by the end of the finger. $C_3$ charges and makes $V_{gs}$ more negative. This increases the channel resistance and thereby decouples less and less of the audio frequency signal so the output signal strength increases. To increase the output signal strength, the capacitor must be discharged by bridging contacts A and B which discharges the capacitor.

The touch volume control can be included in the preamplifier stage (not the power stage) of radio receivers, record players, cassette recorders, and so on.

## 28.4   Touch Switch

The circuit in figure 28.3 is a much better touch switch than the one described in experiment C14 (Section 26.3). The circuit shown uses two VMOS transistors in a bistable circuit. When the finger momentarily bridges the touch contacts, one transistor switches on and the other off thereby switching the relay on or off.

When the power is switched on, capacitor $C_1$ across the gate terminal and the 0 V supply line will ensure that $Tr_2$ is switched off. Consequently, the positive voltage on the drain terminal of $Tr_2$ switches $Tr_1$ on

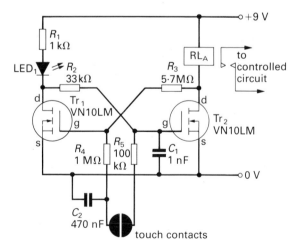

**Figure 28.3** Touch switch

via resistor $R_3$. Thus $LED_1$ lights. This is a stable state and indicates the circuit is ready for action. Now if the touch contacts are bridged by the finger, the positive charge on $C_1$ is transferred to the gate of $Tr_1$ and switches the latter off. The rise in voltage on the drain terminal of $Tr_1$ causes $Tr_2$ to switch on via resistor $R_1$ and to energise the relay. This is also a stable state. These two stable states give the circuit its name *bistable*. Bistables using CMOS logic integrated circuits are described in Book E, Chapter 9.

The circuit will remain in either of the two states until the finger momentarily bridges the contacts. If the finger is left on the contact, the circuit will oscillate giving a low-frequency square wave output, i.e. the relay will alternately switch on and off. The relay may be used to control motors, lamps, etc.

**CAUTION**

Do not use this circuit to control mains-operated loads via the relay unless you are absolutely sure that the circuit is completely isolated from mains voltages.

## 28.5   **Radio Thermometer**

This application outlines an idea for a really miniature device, a very low-power radio transmitter based on a single transistor, $Tr_1$, as shown in figure 28.4(a). The circuit emits pulses of radio waves which are picked up on an ordinary transistor radio. The range of the device is quite small, two to three metres at most. The frequency of the pulses (not of the radio frequency) varies with temperature since the circuit uses a bead thermistor, $Th_1$. It is a fairly simple matter to calibrate the thermometer so that the pulse rate (e.g. in pulses per minute) can be converted into a temperature reading.

The transistor operates as a Hartley oscillator since its emitter is connected one-third along the coil across which is capacitor $C_2$ — it is this part of the circuit

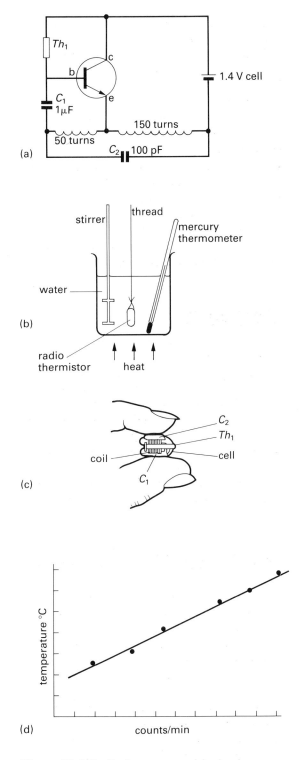

**Figure 28.4** Radio thermometer: (a) circuit (b) assembly of radio thermometer   (c) apparatus for calibration   (d) a calibration graph

which produces radio frequency waves.
When the mercury cell is first connected to
the circuit, capacitor $C_1$ begins to charge at
a rate determined by the resistance of the
thermistor, i.e. by its temperature. When
the rising voltage across $C_1$ has reached
about 0.6 V, $Tr_1$ switches on and the radio
frequency part of the circuit begins to
operate. The frequency of these waves is
determined by the inductance of the coil
and the value of $C_1$. This radio emission
discharges the capacitor and the transistor
switches off. Once again $C_1$ charges and
the next pulse of radio waves is emitted.
The rate at which these pulses are
produced clearly depends on the
temperature of thermistor. Since the
thermistor is an n.t.c. type, the pulse rate
increases as its temperature rises. Use a
1 MΩ resistance glass-bead thermistor,
e.g. type GL16. The pulses can be heard as
a succession of squelchy clicks over most
of the medium and long wave bands of an
AM radio placed a short distance away,
but tune the radio and adjust its position
so that the clicks are heard at their
loudest.

Producing a very small assembly of this
circuit is quite difficult. It is a good idea to
assemble the components on $Tr_1$. Use fine
(not less than 42 s.w.g.) enamelled copper
wire for $L_1$ and wind it close round $Tr_1$ as
shown in figure 28.4(b). The main problem
is to make a springy holder for the
mercury cell so that it can be replaced as
necessary. Do not apply too much heat to
$Tr_1$ and $Th_1$ when soldering the circuit.

Once assembled, the Radio
Thermometer needs calibrating so that the
number of clicks per second can be
interpreted as a temperature. The
calibration can be done with the
arrangement shown in figure 28.4(c) to
produce the calibration graph shown in
figure 28.4(d). Of course, if you don't want
the circuit to work as a radio thermometer,
it can simply be used as a tracer by
replacing $Th_1$ by a fixed value resistor. If
you decrease the value of $C_1$, e.g. to
100 nF, a high pitched whistle will be
heard from the radio.

## 28.6  Intruder Alarm

There are many ways by which unwelcome
visitors can be detected. One of the
simplest electrical methods is to use a trip
wire. The wire need not be electrically
conducting provided a switch is operated
when it is tugged. An alternative method is
to make sure the wire is broken by the
intruder. If the wire carries a small current
which holds a relay on, when the wire is
broken by the intruder, the relay de-
energises thereby switching on an alarm
via a pair of its contacts.

Many of today's intruder alarms use
sophisticated sensing techniques to signal
the presence of an intruder. Movement,
noise, body heat, infrared cameras, and the
breaking of infrared beams are some of the
ways of sensing unwanted entry into
homes and establishments. But this
project describes a simple technique which
uses a small permanent magnet and a reed
switch, $RS_1$ (Book A, Chapter 5) to
indicate when a door or window is open.

The circuit of figure 28.5(a) shows how

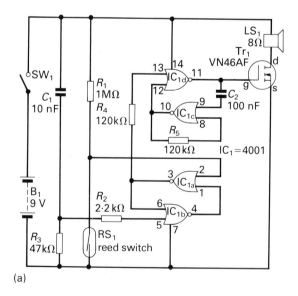

(a)

**Figure 28.5**  Intruder alarm  (a)  circuit

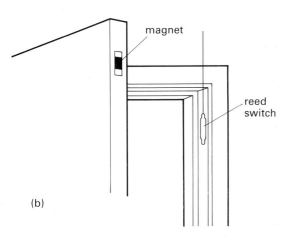

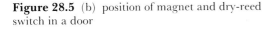

**Figure 28.5** (b) position of magnet and dry-reed switch in a door

easy it is to interface low-power CMOS devices (here a quad 2-input NOR gate, type 4001 — see Chapter 3, Book E) to a high-power VMOS transistor, $Tr_1$. The four CMOS NOR gates in $IC_3$ are wired up as two oscillators. One oscillator, $IC_{1a}$ and $IC_{1b}$, produces a low frequency of about 1 Hz. These on/off pulses switch on and off an 800 Hz audio frequency produced by NOR gates $IC_{1c}$ and $IC_{1d}$. The pulses of audio tone are then amplified by means of the VMOS transistor, $Tr_1$.

However, the oscillators do not work when the reed switch, $RS_1$, is closed by the magnet, i.e. when a door or window is closed, for this holds pin 2 of $IC_{1a}$ at 0 V. When the reed switch is opened, resistor $R_1$ makes pin 2 of $IC_{1a}$ high and the oscillator works. The standby current drain from the power supply is very low until the circuit operates. It is worth altering the values of the two capacitors, $C_1$ and $C_2$, and of the two resistors, $R_4$ and $R_2$, to obtain the most attention-drawing sound.

## 28.7  Radio Receiver

Figure 28.6 shows a simple two-stage radio receiver which is a considerable improvement on the radio receiver described in Chapter 7. The use of the

JUGFET, $Tr_1$, provides a sensitive and highly selective radio receiver. The advantage of using a JUGFET in the first stage of the receiver is that the tuned circuit, comprising $VC_1$ and $L_1$, has a high resistance across it. This means that a carrier wave tuned in by the circuit generates a large voltage across the tuned circuit at the frequency of the carrier wave. The receiver therefore not only separates closely spaced stations, but is also capable of 'bringing in' weak stations.

The circuit will tolerate many different types of bipolar transistor, $Tr_2$, in the second stage. This transistor is designed to both amplify and detect, i.e. rectify, the carrier wave. It may be necessary to adjust the values of $R_1$ and $R_2$ to obtain optimum performance with a particular bipolar transistor. You should wind between 30 and 50 turns of enamelled copper wire round the ferrite rod, and select a 100 pF tuning capacitor. The circuit should work in most areas without an aerial.

A high impedance earpiece is required for this circuit, so use a crystal earpiece or 2000 ohm headphones if you can obtain them. An add-on amplifier (see Book D, Chapter 2) will provide increased audio power for loudspeaker reception. Though this radio receiver is simple and cheap to build, you will find an even better design in Book D, Section 2.5 which is based on a purpose-designed integrated circuit for radio receivers.

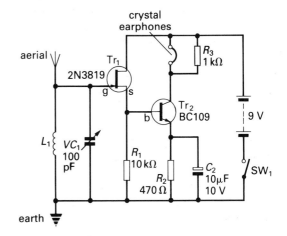

**Figure 28.6** Radio receiver

## 28.8  Thermostat

The circuit shown in figure 28.7 is a better transistor switch than the simple Darlington-pair circuits described in Chapter 20. The main problem with simple transistor switches like the Darlington pair is that the switch goes on and off at one particular voltage. This behaviour can cause problems at the point when the circuit is switching a relay on and off. If electrical noise, e.g. mains pick-up, is present, along with the steady d.c. signal that operates the Schmitt trigger, the relay will 'chatter' when it is energising and de-energising.

A Schmitt trigger is a sharp electronic switch which overcomes this problem. It has the effect of making the switch-on temperature slightly higher than the switch-off temperature — a function which is called hysteresis. The result is a thermostat, a heat operated electronic switch, which snaps on and snaps off at two slightly different temperatures. The way a transistorised Schmitt trigger works is described in Section 25.5. Schmitt triggers can be obtained in integrated circuit packages as described in Books D and E.

A thermistor, $Th_1$, should be chosen which has a resistance not less than 4.7 k$\Omega$ (at 25°C). The value of $VR_1$ should then be selected to enable the circuit to switch at

the desired temperature. The transistors are not critical so use any general purpose npn and pnp types which have a d.c. current gain of at least 100. Make sure that $Tr_3$ is able to carry the current required by the relay. Use a relay whose contacts are capable of switching the heater power; if this is a mains heater make sure the relay is designed for mains operation.

**CAUTION**

Do not use a mains heater with a relay unless you are quite clear how to make the appropriate earth connection to the heater and the thermostat you have designed.

Consider using the thermostat for the following purposes:

(i)  environmental control in greenhouses, animal nurseries, reptile ponds, and tropical fish aquaria;
(ii)  the control of central heating systems;
(iii)  fire warning.

Consider modifying the thermostat circuit so that it is triggered by a change of light intensity, or of dampness, or of magnetic field strength.

## 28.9  Transistor Tester

The circuit shown in figure 28.8 makes a very simple transistor tester. The circuit is

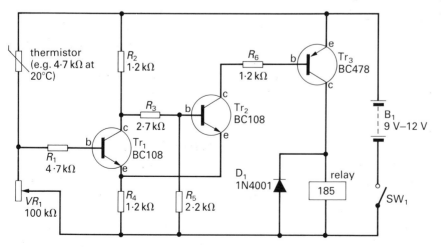

**Figure 28.7**  Thermostat

set up using a complementary pair of transistors, $Tr_1$ and $Tr_2$, which are known to work. These transistors may be any general-purpose npn and pnp types which temporarily plug into transistor sockets. The value of $R_1$ (and also possibly $C_1$) is adjusted so that the loudspeaker produces an audio tone — between a buzz and a whistle!

Now suppose you want to test a suspect npn transistor? First unplug the npn transistor which works in the circuit and replace it by the possible faulty transistor. If the circuit continues to produce a tone, the transistor is working. If the oscillations have ceased, the transistor is faulty. You test a pnp transistor in a similar way by substituting $Tr_2$ with the transistor being tested.

If you don't know what the transistor pin connections are, or whether it is an npn or pnp type, simply plug the transistor in the sockets in all possible configurations. If you don't hear a buzz after all that effort, the transistor tested must be faulty!

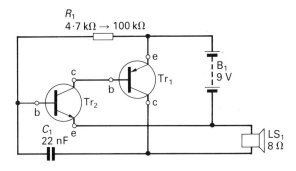

**Figure 28.8**  Transistor tester

## 28.10  **Blind Aid**

The simple two-transistor oscillator shown in figure 28.9 is designed to produce an audible tone which varies with light intensity. The tone rises in pitch as the light-dependent resistor, $LDR_1$, becomes more strongly illuminated. With the values shown, the circuit produces a low frequency clicking in the dark which rises to a high-pitched note in the light. You

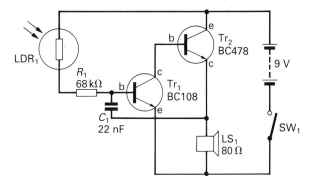

**Figure 28.9**  Blind aid

may need to experiment with the values of $R_1$ and $C_1$ to produce the frequency range you want: reducing the values of $R_1$ and $C_1$ increases the frequency.

The circuit has a number of applications:

**(a)**  Blind people need to know whether they are wearing light or dark clothes; whether it is day or night; whether a light is on in a room or not.

**(b)**  As an aid to allow blind people to use a kitchen balance, or other household appliances. Can you think how?

**(c)**  Flights of stairs, open doors, and other obstacles are hazards to blind people. Think of ways in which the circuit could help them overcome these difficulties.

Talk to blind people about their problems and see whether a modification of this circuit could be of use to them.

Use a 1 M$\Omega$ thermistor, e.g. the bead thermistor type GL16, in place of $LDR_1$ to make the circuit responsive to temperature change. How could you use the circuit to indicate moisture differences?

## 28.11  **Ghost Finder**

The circuit shown in figure 28.10 produces an audio frequency tone which varies with the strength of an electric field in the vicinity of the metal plate. Aren't ghosts, unseen or spectral, supposed to change the electric field in the space around them?

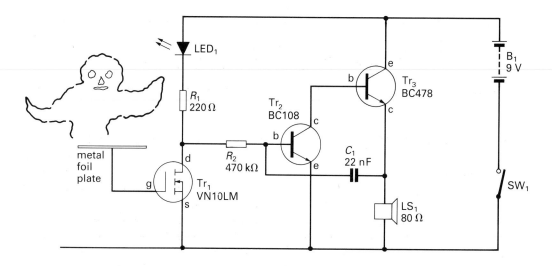

**Figure 28.10** Ghost detector

Ghosts have kept away from my house since I built the circuit, but when I comb my hair the circuit 'wails' and signals the changing electric field produced by static electricity on the comb. Ghost hunting apart, the circuit could be used to investigate electric fields which is what an electroscope does. The circuit might well predict the onset of thundery weather. See what use you can make of it.

The circuit demonstrates the high input resistance of a VMOS transistor, $Tr_1$. The drain-to-source resistance of $Tr_1$ changes with the gate-to-source voltage (Section 26.5), and negligible current flows into the gate terminal to make this happen. Now the gate-to-source voltage is provided by the electric field in the region of the metal plate. When a ghost enters a room, the electric field changes which varies the drain-to-source current flowing through $Tr_1$. This current varies the brightness of $LED_1$.

The two-transistor oscillator based on $Tr_2$ and $Tr_3$ provides an audio tone which changes with the strength of the electric field. The rate at which capacitor $C_1$ charges and discharges via resistor $R_2$ is determined by the voltage at the drain terminal of $Tr_1$. Therefore the audio tone varies. You may need to experiment with the values of $R_2$ and $C_1$ to obtain a good variation of frequency. Check the operation of the circuit by combing your hair and bringing the comb near the metal plate. Use metal foil, brass or copper to make a 100 mm diameter, or larger, plate. Good hunting!

# 29 Project Modules

## 29.1 **What they are**

At the end of each book of *Basic Electronics* there are a number of practical projects for you to build. These projects are called Project Modules and there are thirty five of them in all. This chapter describes how to build the seven Project Modules shown in figure 29.1. They are:

C1: Voltage Booster
C2: Geiger Counter
C3: Decade Counter
C4: Thermometer
C5: Bipolar Transistors
C6: Comparator
C7: Strain Meter

The Project Modules enable you to build up a set of electronic building blocks which can be connected together in various ways to design useful and interesting electronic systems. Details are provided for assembling each Project Module on a printed circuit board (PCB), and for interconnecting it with other project modules using flying leads.

Before assembling the circuits, you should read Book A, Section 6.3, which gives guidance on the preparation of the PCBs. You should also read Book A, Section 12.2 which gives hints on handling the CMOS devices used in Project Module C3. The references to other parts of *Basic Electronics* provide further information of the devices and circuits used in the Project Modules. The examples for using the Project Modules to design electronic systems should enable you to solve other problems which wholly or in part have an electronic solution.

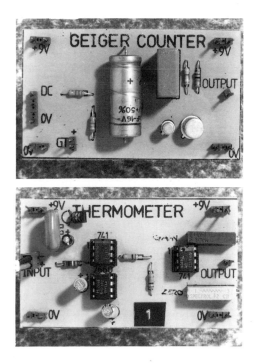

**Figure 29.1** Project Modules C1, C2, C3 and C4

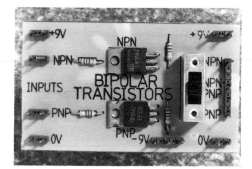

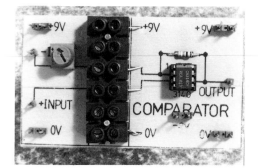

**Figure 29.1**  Project Modules C5, C6 and C7

## 29.2 *Project Module* C1

### Voltage Booster

This module provides about 200 V a.c. and about 400 V d.c. from a 9 V battery. It needs to be driven with the rectangular waveform obtained from the Astable (Project Module A6). The 400 V d.c. voltage is suitable for operating the Geiger Counter (Project Module C2) for the detection of nuclear radiation. The 250 V a.c. is suitable for operating small fluorescent tubes.

**CAUTION**

The d.c. voltage produced by the Voltage Booster can be as high as 500 V. However, the current it generates is very small. Therefore it will give you an unpleasant but harmless electrical shock if you should touch its output terminal. Do not use the Voltage Booster for any purpose other than that described below.

### *Circuits*

Figure 29.2 shows how to generate a high voltage by feeding pulses from the Astable into the low voltage windings of a mains transformer, $T_1$, which is used as a step-up transformer. The pulsing high voltage is rectified and doubled by the two diodes, $D_1$ and $D_2$, and the two capacitors, $C_2$ and $C_3$, to give a d.c. voltage of about 400 V. Chapter 10, Book C, explains how this 'voltage doubler' works.

It is easy to adjust the d.c. voltage by varying the frequency of the Astable which has the effect of 'tuning' the transformer. Capacitor, $C_1$, couples the signal from the Astable to the low-voltage windings of the transformer. Note that capacitors $C_2$ and $C_3$ must have a working voltage of at least 630 V.

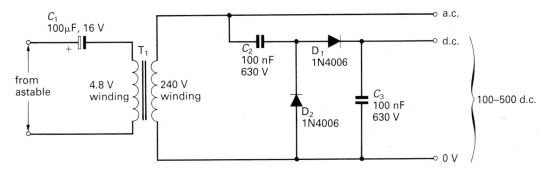

**Figure 29.2** Circuit diagram of the Voltage Booster

## Components and materials

$T_1$: low-voltage mains transformer, PCB
  mounting type having two secondary
  windings each of 4.5 V only one of which
  is used.

$C_1$: electrolytic capacitor, value 100 $\mu$F,
  16 V working

$C_2$, $C_3$: polycarbonate capacitor, 100 nF,
  630 V working

battery: 9 V PP9

battery clip: PP9 type

wire: multistrand, e.g. 7 × 0.2 mm; single
  strand; e.g. 1 × 0.6 mm interconnections
  between other modules

PCB: 90 mm × 50 mm

connectors: PCB header and PCB socket
  housing; crimp terminals

$D_1$, $D_2$: silicon rectifier diodes, type
  1N4006

## PCB assembly

Figure 29.3 shows the layout of the
components on the PCB, and figure 29.4
the copper track pattern on the other side
of the PCB. See Section 6.3, Book A, for
guidance on the preparation of the PCB.

Make sure that the diodes, $D_1$ and $D_2$,
and the electrolytic capacitor, $C_1$, are
soldered to the PCB the right way round.
The polycarbonate capacitors, $C_2$ and $C_3$,
may be connected either way round. The
terminal pins for the connections on the
PCB are made by cutting single and double
pins from the PCB header. Single sections
of the PCB socket housing are used for
terminating wire ends for connecting the
Voltage Doubler to the battery and to other
Project Modules. Strip 5 mm of insulation
from the ends of the wires, and use a

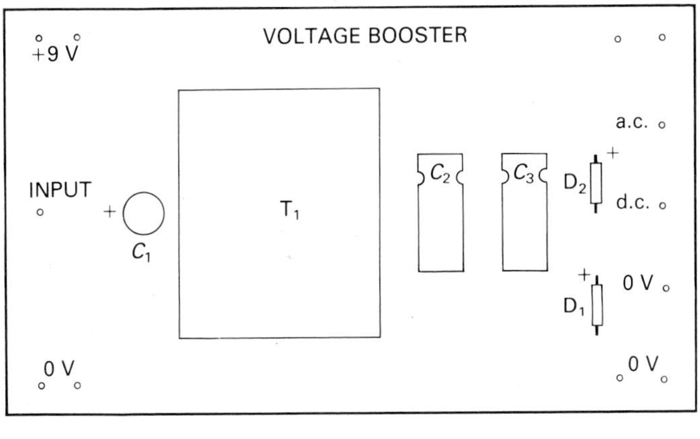

**Figure 29.3** Component layout of the PCB (actual size)

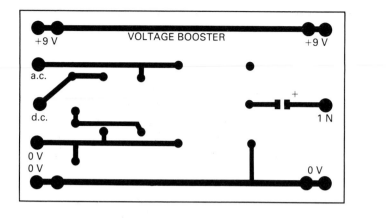

**Figure 29.4** Track pattern for the PCB (actual size)

crimping tool to squeeze a crimp connector on the bare ends. Push the crimp connector into the PCB socket until it clicks into place.

## Testing

Wire the following components to the Astable (Project Module A6): $R_1 = 4.7$ k$\Omega$; $R_2 = VR_1$, a 100 k$\Omega$ variable resistor; $C_1 = 220$ nF. Verify that the Astable works by connecting the Piezo-sounder to its output. Note the chage of frequency when $VR_1$ is adjusted. This signal is used to 'tune' the transformer and very the voltage produced by the Voltage Booster.

Connect a high resistance voltmeter, e.g. a digital voltmeter, to the d.c. output of the Voltage Booster. Connect the output of the Astable to the input of the Voltage Booster, and the two 0 V connections together as shown in figure 29.5. Adjust the variable resistor, $VR_1$, on the Astable and note that the d.c. output of the Voltage Booster can be varied. The maximum voltage obtained should be about 450 V d.c. Experiment with the values of $R_1$, $VR_1$ and $C_1$ in the Astable if the d.c. voltage does not exceed 450 V.

## Using the Voltage Booster

The Voltage Booster is intended to operate the Geiger Counter (Project Module C2) which needs about 400 V d.c. (depending on the Geiger tube used). You might like to make up a portable battery-operated lamp by using the a.c. voltage output from the Voltage Booster to light a small fluorescent tube — the type used in caravans and some lanterns.

### CAUTION

Do not attempt to light a mains-operated fluorescent tube while the tube is in its mounting.

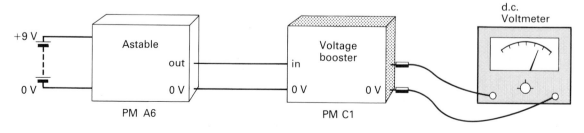

**Figure 29.5** Testing the Voltage Booster

# 29.3 *Project Module* C2

## Geiger Counter

### *What it does*

This Project Module detects nuclear radiation, especially beta particles and gamma rays, and also the background radiation produced by cosmic rays and the radioactive substances in materials around us, such as uranium bearing rocks. The main component of the Geiger Counter is a Geiger tube. For every beta particle or gamma ray quantum that enters the Geiger tube, a 'click' is produced in the Piezo-sounder (Project Module A3) connected to it. By feeding the pulses that make the clicks to the Two-Digit counter (Project Module D7), the number of particles can be counted.

### *Circuit*

Figure 29.6 shows that the Geiger counter requires two d.c. power supplies. 500 V is needed to operate the Geiger tube, and 9 V for the two-transistor amplifier. The 500 V d.c. supply is obtained from the Voltage Booster (Project Module B7), and the 9 V supply from a battery. Each time the residual gas inside the Geiger tube is ionised by the passage through it of an ionising particle or quantum of radiation,

a brief change of voltage occurs across resistors, $R_1$ and $R_2$. Capacitor $C_2$ couples this voltage pulse to the two-transistor Darlington-pair amplifier, $Tr_1$ and $Tr_2$, so that each pulse gives a sharp click in the Piezo-sounder. Since capacitor $C_2$ has a large voltage difference across it, it must have a working voltage of at least 500 V otherwise it will break down. The Darlington pair transistor amplifier is discussed in Chapter 20.

### *Components and materials*

$Tr_1$, $Tr_2$: npn transistors, BC108 or similar
$R_1$ to $R_4$: fixed-value resistors, values
   2.2 M$\Omega$, 2.2 M$\Omega$, 470 k$\Omega$, and 1 k$\Omega$,
   respectively, 0.25 W, $\pm$ 5%
$C_1$: electrolytic capacitor, 1000 $\mu$F, 16 V
   working
$C_2$: polycarbonate capacitor, 100 nF,
   630 V working
$D_1$: zener diode, 4.7 V, 400 mW
battery: 9 V PP9
battery clip: PP9 type
wire: multistrand, e.g. 7 × 0.2 mm; single
strand: e.g. 1 × 0.6 mm
Geiger tube: any low-voltage type which
   has a threshold of about 400 V
PCB: 90 mm × 50 mm
connectors: PCB header and PCB socket
   housing; crimp terminals

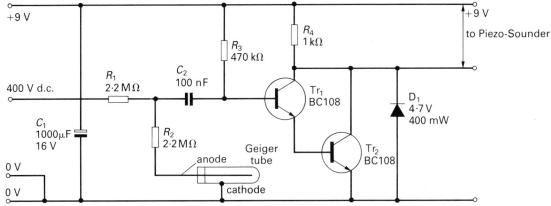

**Figure 29.6** Circuit diagram of the Geiger Counter

## PCB assembly

Figure 29.7 shows the layout of the components on the PCB, and figure 29.8 the copper track pattern on the other side of the PCB. See Book A, Section 6.3, for guidance on the preparation of the PCB.

The anode and cathode terminals on the Geiger tube must be connected to the circuit the right way round, i.e. the cathode to the 0 V, and the anode to the positive of the high voltage d.c. supply. Geiger tubes are generally made of glass, and they are also expensive, so they should be handled with care and protected from sharp knocks.

The terminal pins for the connections on the PCB are made by cutting single and double pins from the PCB header. Single

sections of the PCB socket housing are used for terminating wire ends for connecting the Geiger Counter to the battery and to other Project Modules. Strip 5 mm of insulation from the ends of the wires, and use a crimping tool to squeeze a crimp connector on the bare ends. Push the crimp connector into the PCB socket until it clicks into place.

## Testing and use

Connect the Geiger Counter to the Astable (Project Module B6) and the Voltage Booster (Project Module C1) as shown in figure 29.9. Find the correct operating voltage of the Geiger tube you are using. This voltage is usually the threshold voltage and is the lowest voltage at which the tube will detect ionising radiations.

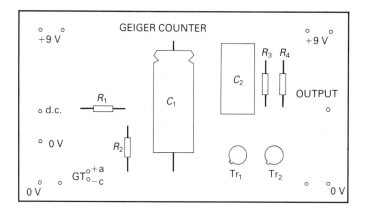

**Figure 29.7** Component layout on the PCB (actual size)

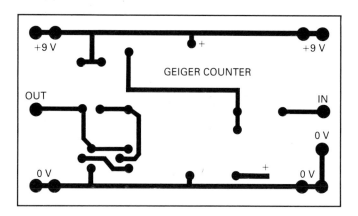

**Figure 29.8** Track pattern on the PCB (actual size)

The tube will operate over a range, e.g. 75 V, of voltage called the 'plateau' of the tube. The sensitivity of the tube is roughly constant over the plateau, and it is usual to set the operating voltage about half way along the plateau.

Set a high resistance voltmeter to 500 V or 1000 V d.c. and measure the output voltage of the voltage booster. Adjust $VR_1$ on the Astable so that the d.c. voltage is roughly half way along the plateau of the tube. Connect the Geiger Counter to the Voltage Booster, and the Piezo-sounder to the output of the Geiger Counter. You should soon hear random clicks from the Piezo-sounder as the Geiger tube responds to background radiation. Make sure that the Geiger tube does not operate above the voltage of the plateau since this will make it go into continuous discharge and damage it.

How can you use the Two-Digit Counter (Project Module D7) to count the number of particles detected?

Could you use the Monostable (Project Module B4) to count the number of particles in a given time?

The following program for the BBC microcomputer plots a bar graph display of the counts detected by the Geiger Counter. The computer must be fitted with a suitable input/output interface such as the one available from Technology Teaching Systems Ltd, Penmore House, Hasland Road, Hasland, Chesterfield, S41 0SJ. This interface is supplied in kit form and comes complete with cables for connecting it to the user and printer ports

of the BBC microcomputer. It provides eight binary-weighted input lines and eight binary-weighted output lines which can be switched on and off individually by the microcomputer. LEDs on the interface signal when an output is at logic 1, and 4 mm sockets enable data to be fed to it or taken from it. The pulses from the Geiger counter are fed to bit 2 of the input sockets on the interface where they are loaded into the computer's internal 16-bit counter for processing. Note that the 0 V supply of the Geiger counter is connected to the 0 V connection on the interface.

Nine procedures at lines 90, 160, 270, 330, 430, 510, 550, 680, and 760, control this program which plots a ten-stage bar graph indicating the counts accumulated in the chosen period of time, and prints the count rate and the average count.

The selection of sampling rate and vertical scaling of the graph is achieved by lines 270 to 320.

Lines 330 to 420 print numbers 1 to 9 along the bottom of the screen at the graphics cursor for more accurate positioning of these numbers.
Lines 430 to 500 define the window at the top of the screen in which data is printed.
Lines 510 to 540 print the present count rate and the average count rate in the graphics window. Note that after each sample, numbers only are printed in the gaps to increase computing speed and to prevent flicker.
Lines 550 to 670 count the pulses received via bit 2 of the input port. Lines 580 and

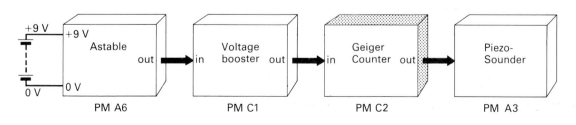

**Figure 29.9** System for the Geiger Counter

590 set the BBC's internal counter to 65535. Line 600 provides the chosen delay between sampling. Line 610 then finds the number left in the internal counter which holds a 16-bit number. Line 620 works out the number of counts. Lines 680 to 750 plot the result.

Line 640 updates the total count. Lines 770 to 790 print the results in the graphics window.

Lines 90 to 150 ask the user whether he or she wants the sampling to be repeated.

```
 10  REM "GEIGER COUNTER"
 20  PROCset_up
 30  MODE5
 40  PROCstart_screen
 50  MODE1
 60  PROCbar_numbers
 70  PROCwindows
 80  PROCprint_results_mask
 90  PROCcount
100  VDU7
110  PRINTTAB(1,3);"Do You Want To Repeat This?'Y' or 'N'"
120  A$=GET$
130  IF A$="Y"GOTO20
140  IF A$="N"THEN PRINTTAB(1,3);"Run this program again if you want to"
150  END
160  DEFPROCset_up
170  total%=0
180  count%=65128
190  acr%=65131
200  pcr%=65132
210  ddr%=65122
220  ?ddr%=0
230  ?acr%=32
240  ?pcr%=0
250  @%=1
260  ENDPROC
270  DEFPROCstart_screen
280  PRINTTAB(3,2);"GEIGER COUNTER";TAB(2,7);"HOW MANY SECONDS";
       TAB(2,9); "BETWEEN SAMPLES?"
290  INPUThowlong
300  PRINTTAB(1,15);"WHAT SCALE FACTOR?"
310  INPUTscale%
320  ENDPROC
330  DEFPROCbar_numbers
340  VDU5
350  MOVE17,45
360  FOR N=1 TO 9:PRINTN;
370  PLOT0,105,0
380  NEXTN
390  PLOT0,-25,0
400  PRINT10
410  VDU4
```

```
420 ENDPROC
430 DEFPROCwindows
440 VDU24,0;85;1279;850;
450 VDU28,0,5,39,1
460 COLOUR129
470 CLS
480 @%=&20205
490 MOVE0,85
500 ENDPROC
510 DEFPROCprint_results_mask
520 PRINTTAB(1,0);"Present Count Rate =            ";"counts/sec"
530 PRINTTAB(1,1); "Average Count Rate =            ";"counts/sec"
540 ENDPROC
550 DEFPROCcount
560 FORsample%=1 TO 10
570 TIME=0
580 ?count%=255
590 ?(count%+1)=255
600 REPEAT:UNTIL TIME> howlong*100
610 result%=?(count%+1)*256+?count%
620 result%=65535-result%
630 PROCbar(result%*scale%)
640 total%=total%+result%
650 PROCprint_results
660 NEXTsample%
670 ENDPROC
680 DEFPROCbar(high%)
690 PLOT80,0,high%
700 PLOT80,60,0
710 PLOT81,0,-high%
720 PLOT80,-60,0
730 PLOT81,0,high%
740 PLOT80,136,-high%
750 ENDPROC
760 DEFPROCprint_results
770 PRINTTAB(22,0)result%/howlong
780 PRINTTAB(22,1)total%/sample%/howlong
790 ENDPROC
```

# 29.4 *Project Module* C3

## Decade Counter

### *What it does*

This Project Module has ten outputs which go high one after the other as it is fed with a regular series of input pulses. Only one output is high at any time. The high (i.e. a voltage near the supply voltage) at an output can be used to switch on a motor, lamp, etc., by being connected to an input of the Relay Driver (Project Module B2). The rate at which the ten outputs switch on and off is determined by the frequency of a 'clock', e.g. the Astable (Project Module A6), connected to the input of the Decade Counter. The Decade Counter can be reset to zero (i.e. to output '0') at any count in its cycle from 0 to 9. Also its counting can be disabled at any count so that an output is held high while the counter ignores input pulses.

### *Circuit*

The circuit shown in figure 29.10 is based on a single CMOS integrated circuit, $IC_1$, type 4017. Note that its ten output pins are not numbered in sequence round the package. $IC_1$ has an input (pin 14) to which are fed CLOCK pulses, and a RESET terminal (pin 15). A divide-by-ten output (pin 12) enables the 4017 to operate a second counter so that every ten input pulses can be counted, i.e. two cascaded 4017s could be used to count to 99. The state of an output is shown by the ten light emitting diodes, $LED_0$ to $LED_9$. The resistors $R_3$ to $R_{12}$ limit the current flowing through these light emitting diodes, and allow the 4017 to produce a large 'swing' of voltage at each output so that other project modules can be reliably operated. The push-to-make, release-to-break switches $SW_1$ and $SW_2$ are the RESET and DISABLE switches,

respectively. The 4017 decade counter, which is one of a number of ICs that are also called frequency dividers, is described in Book E, Chapter 6.

### *Components and materials*

$IC_1$: decade counter, CMOS type 4017
IC holder: 16-way
$R_1$ to $R_{11}$: fixed-value resistors, values
    47 kΩ, 47 kΩ, 1.5 kΩ ($R_3$) 0.25 W, ± 5%
$LED_0$ to $LED_9$: light emitting diodes
$SW_1$, $SW_2$: push-to-make, release-to-break
    switches
battery: 9 V PP9
battery clip: PP9 type
wire: multistrand, e.g. 7 × 0.2 mm; single
strand: e.g. 1 × 0.6 mm
PCB: 90 mm × 50 mm
connectors: PCB header and PCB socket
    housing; crimp terminals

### *PCB assembly*

Figure 29.11 shows the layout of the components on the PCB, and figure 29.12 the copper track pattern on the other side of the PCB. See Book A, Section 6.3, for guidance on the preparation of the PCB.

Make sure that $LED_0$ to $LED_9$ are connected the right way round; all the cathode terminals are to face the lower edge of the board. The terminal pins for the connections on the PCB are made by cutting single and double pins from the PCB header. Single sections of the PCB socket housing are used for terminating wire ends for connecting the Decade Counter to the battery and to other Project Modules. Strip 5 mm of insulation from the ends of the wires, and use a crimping tool to squeeze a crimp connector on the bare ends. Push the crimp connector into the PCB socket until it clicks into place.

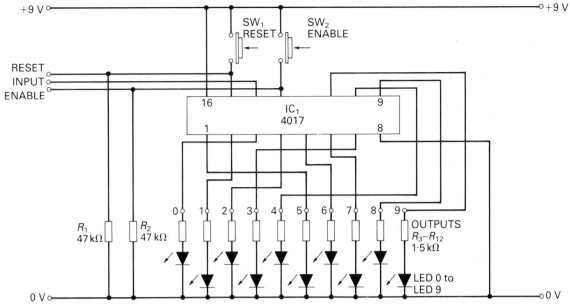

**Figure 29.10**  Circuit diagram of the Decade Counter

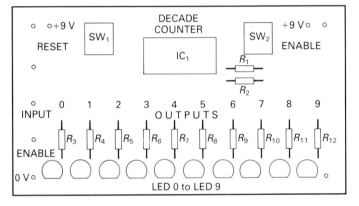

**Figure 29.11**  Component layout on the PCB (actual size)

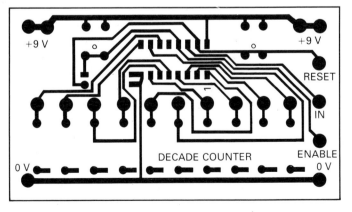

**Figure 29.12**  Track layout on the PCB (actual size)

## Testing and use

Wire up the pulser (Project Module A2) so that $R_1$ is replaced by a 100 kΩ variable resistor, $VR_1$. Capacitor $C_1$ should have a value of 10 μF. Connect a 9 V battery to the Pulser and adjust $VR_1$ so that $LED_1$ on the Pulser flashes on and off about once per second (1 Hz).

Connect the output of the Pulser to the input (CLOCK) terminal of the Decade Counter as shown in figure 29.13. Don't forget to make the power supply connections between the modules. The light emitting diodes on the Decade Counter will light one after the other, counting '0' to '9' and recycling. The rate at which they light can be controlled by $VR_1$ on the Pulser.

While the LEDs are switching on and off in sequence, press $SW_1$ and you will notice that this resets the count to '0'. Release $SW_1$ and the counting begins again. Next press $SW_2$ and you will notice that count will 'freeze' at whatever output was high just before $SW_2$ is pressed. Release $SW_2$ and the counting continues.

Next link pin 1 (the RESET terminal) to output 6, i.e. $LED_6$. You will notice that the count goes from '0' to '5' and returns to '0' automatically. Once $LED_6$ attempts to come on, the rising voltage at output 6 resets the counter via the signal fed back to the RESET terminal.

It is simple to connect the Relay Driver (Project Module B2) to the output of the Decade Counter. You can select two different outputs from the Decade Counter so that the two relays are energised at different times. Up to ten relays can be controlled in sequence this way. Each relay could be used to switch on a lamp, electric motor, or some other device at regular intervals.

## Questions

2  How would you use the Monostable (Project Module B4) with the Decade Counter so that once a relay has been energised, it remains on for a period of time determined by the product of the capacitor and resistor values on the monostable? (See Book B, Section 22.5).

3  How would you use the Decade Counter to switch on a lamp after one hour? Once switched on, the lamp has to stay on for one minute before switching off.

4  Explore ways of including the Bistable (Project Module B3) with the Decade Counter and Relay Driver for counting and control applications.

5  Use the Decade Counter to count the pulses delivered by the Geiger Counter (Project Module C2). Use two Decade Counters to count up to 99.

## Question

1  How would you use the Decade Counter as (a) an 'electronic die'? (b) a 'reaction timer'?

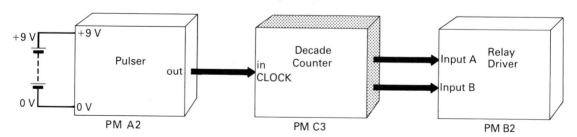

**Figure 29.13** System for controlling the Relay Driver using the Decade Counter

# 29.5 *Project Module* C4

## Thermometer

### *What it does*

This Project Module measures temperature in the range − 50 to 150°C with an accuracy of ± 1°C. The temperature is read on a 1 V f.s.d. moving coil voltmeter, e.g. the Voltmeter (Project Module A4) or a digital voltmeter. The thermometer is easily calibrated using the 'zero' and 'range' controls, $VR_1$ and $VR_2$, respectively, which allow the thermometer to measure, for example, the temperature range 25°C to 45°C.

### *Circuit*

The circuit shown in figure 29.14 uses an operational amplifier, $IC_3$, to provide a constant flow of current through the sensor, $S_1$, which is the base-emitter junction of an npn transistor. The voltage across the base-emitter junction of the transistor is proportional to its temperature. The transistor used in this way makes a low-cost temperature sensor. A silicon diode could be used instead of the transistor. The small variation in voltage across the base-emitter junction is about − 2 mV C$^{-1}$, and has to be amplified by a second op amp, $IC_4$, before the temperature is displayed on the meter. The variable resistor, $VR_1$, is used to set the zero reading on the meter, and $VR_2$ the range of the temperature measurement.

Op amps, $IC_3$ and $IC_4$, have to operate from a stabilised power supply of + 5 V (on pins 7) and − 5 V (on pin 4) if the readings on the meter are to be repeatable and accurate. These two voltages are stabilised by the voltage regulators $IC_1$ (+ 5 V) and $IC_2$ (− 5 V). Capacitors $C_1$ to $C_4$ ensure the stable operation of these integrated circuit voltage regulators. Op amps are described in Book D, and voltage regulators of the type used in the Thermometer are discussed in Chapter 9.

### *Components and materials*

$IC_1$: + 5 V voltage regulator, type 78L05
$IC_2$: + 5 V to − 5 V voltage converter, type 7660
$IC_3$, $IC_4$: op amp, type 741

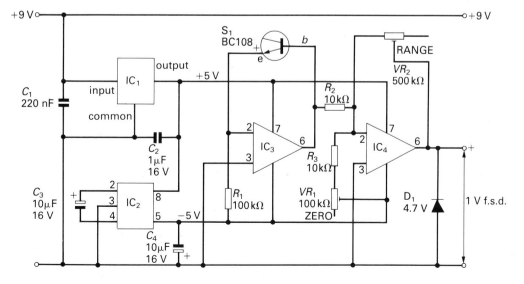

**Figure 29.14** Circuit diagram of the Thermometer

$D_1$: Zener diode, value 2.7 V, 400 mW

IC holders: 2 × 8-way

$C_1$: polyester capacitor, value 220 nF

$C_2$ to $C_4$: electrolytic capacitors, values
1 μF, 10 μF, and 10 μF, respectively,
16 V working

$R_1$ to $R_3$: fixed-value resistors, values
100 kΩ, 10 kΩ, and 10 kΩ,
respectively, 0.25 W, ± 5%

$VR_1$, $VR_2$: 0.75 inch (19.1 mm) 20-turn
variable resistor

$S_1$: npn transistor, type BC108

battery: PP9

battery clip: PP9 type

wire: multistrand, e.g. 7 × 0.2 mm

PCB: 90 mm × 50 mm

connectors: PCB header and PCB socket
housing; crimp terminals

## PCB assembly

Figure 29.15 shows the layout of the components on the PCB, and figure 29.16 the copper track pattern on the other side of the PCB. See Book A, Section 6.3, for guidance on the preparation of the PCB.

The terminal pins for power supply and input and output connections are made by cutting single and double pins from a length of PCB connector plug. Similarly, slice off single sections of PCB connector socket to make the leads for connecting the Thermometer to the Voltmeter (Project Module A4).

Prepare the sensor, $S_1$, as shown in figure 29.17. Cut off the collector lead of

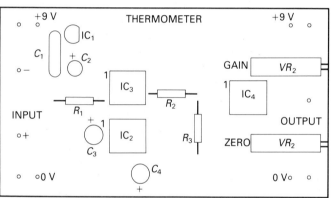

**Figure 29.15** Component layout on the PCB (actual size)

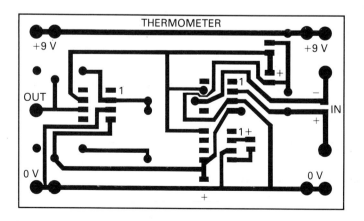

**Figure 29.16** Track pattern on the PCB (actual size)

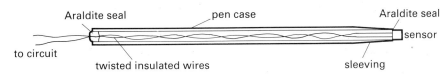

**Figure 29.17** Preparation of the temperature sensor

the transistor close to the can. Solder insulated wires to the base and emitter leads. Insulate the leads from each other and thread them through an empty plastics pen case so that the transistor fits tightly into the tapered end. Glue the transistor in place and also close the other end of the pen case with Araldite to make sure that moisture cannot reach the leads of the transistor. Terminate the leads with PCB connectors so that they can be plugged into the PCB pins.

## Testing and use

Connect the Thermometer to the Voltmeter (Project Module A4), as shown in figure 29.18, and set the meter to its 1 V d.c. range, or use a multimeter. Place the sensor in some crushed ice at 0°C, and adjust $VR_1$ (zero set) so that the meter reads zero. Next place the sensor in boiling water which, if you are close to sea level, should be at 100°C. Adjust $VR_2$ (gain set) so that the meter reads fulls scale. Put the sensor back in the melting ice and make any adjustment necessary to the zero reading. Make one more adjustment to the upper end of the scale and the Thermometer should be ready to use. It is easy to interpret the 0 V to 1 V readings on the meter in degrees Celsius since they will be linear, i.e. 0.5 V reads 50°C. The thermometer can be set to measure a smaller or larger range if required. But note that the sensor should not be used above 150°C or below − 50°C.

If your computer is fitted with an analogue-to-digital converter (ADC) (Chapter 10, Book D), the readings from the Thermometer can be displayed on a VDU. The two Basic programs below are written for the BBC microcomputer which has an analogue port for reading changing voltages. A suitable connector unit that plugs into this port can be obtained from Technology Teaching Systems Ltd, Penmore House, Hasland Road, Hasland, Chesterfield, S41 0SJ. This unit is supplied in kit form. The four analogue channels of the BBC microcomputer are accessible through four 4 mm sockets on the connector unit. The BBC's 1.8 V reference and 5 V supply voltages are also available on this unit.

The Zener diode, $D_1$, in figure 29.15 ensures that the BBC microcomputer cannot be damaged by excessive voltage fed to its analogue port. Note also that the Thermometer should be operated from its own 9 V d.c. supply. A PP9 battery is very suitable. The output of the Thermometer should be connected to channel 0, i.e. ADVAL(1), of the connector unit, and the 0 V supply connected to the 0 V plug on the connector unit.

## Program 1 Max/Min Thermometer

This simple program displays the current, maximum and minimum temperatures reached in a period of time.

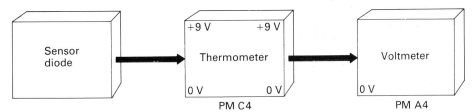

**Figure 29.18** System for the Thermometer

Lines 30 to 60 display the three temperatures.

Line 70 assigns a variable to the output voltage from the Thermometer which is accessed via channel 1 of the analogue port. Division by 360 converts the analogue reading to a number equal to temperature.

Line 100 prints the initial value of A at all three positions of the temperature display.

Line 110 introduces a one second time delay between readings.

Line 120 takes a new value of the input voltage representing temperature.

Line 130 decides whether or not the temperature has changed; if it has, it directs the program to a procedure at line 140 to update the maximum and minimum temperatures, otherwise the program loops back to line 100 and continues to display an unchanged result. The program continues indefinitely between lines 90 and 180.

**Program 1**

```
 10 REM "MAX/MIN THERMOMETER"
 20 CLS
 30 PRINTTAB(1,6);"MAX/MIN THERMOMETER"
 40 PRINTTAB(4,12); "NOW TEMP=      C"
 50 PRINTTAB(4,14); "MAX TEMP=      C"
 60 PRINTTAB(4,16); "MIN TEMP=      C"
 70 A=INT(ADVAL(1)/360)
 80 P=A:R=A:Q=A
 90 REPEAT
100 PRINTTAB(13,12);P;TAB(13,14);Q;TAB(13,16);R
110 I=INKEY(100)
120 X=INT(ADVAL(1)/360)
130 IF X◊P THEN PROCmaxmin ELSE 100
140 DEFPROCmaxmin
150 IF X›Q THEN Q=X ELSE Q=Q
160 IF X‹R THEN R=X ELSE R=R
170 P=X
180 UNTIL FALSE
```

## Program 2 Temp/Time Plot

This program produces a graph of temperature against time.

Line 20 selects a graphics mode that displays large-size printing on the VDU. Lines 30, 40 and 50 identify three procedures that structure the program. The procedure at line 60 designs a graphics window for the graph and prints labels for the graph's axes.

The procedure at line 120 draws in the axes of the graph.

The procedure at line 160 plots the graph after the user has decided (lines 170, 180 and 260) what interval of time is needed between readings.

Lines 210 and 220 select the reading from channel 1 of the analogue port and converts it to a number equal to temperature.

Line 230 prints the value of the temperature in the graphics window.

Lines 240 and 250 plot points on the graph.

When plotting is complete, line 270 selects a procedure that asks whether readings are to be repeated.

**Program 2**

```
 10 REM "TEMP/TIME PLOT"
 20 MODE5
 30 PROCwindow
 40 PROCaxes
 50 PROCplot
 60 DEFPROCwindow
 70 CLS
 80 VDU 24,30;400;1179;1000;:GCOL 0, 130
 90 CLG
100 VDU5:GCOL0,0:MOVE55,940:PRINT"t":MOVE55,880:PRINT"e":MOVE55,820:
    PRINT"m":MOVE 55,760:PRINT"p":MOVE860,450:PRINT"time" :GCOL0,1:MOVE
    350,950:PRINT"TEMP/TIME":VDU4
110 ENDPROC
120 DEFPROCaxes
130 GCOL0,0:MOVE 120, 460:DRAW 1080,460:MOVE 120, 460:DRAW 120, 960
140 FORX=240 TO 1080 STEP 120:MOVE X,460:DRAWX,470:NEXTX
150 ENDPROC
160 DEFPROCplot
170 PRINTTAB(0,25);"PLOT RATE?:ENTER SEC"
180 INPUT S
190 F=130
200 REPEAT
210 CH0= ADVAL(1)
220 N=CH0/360
230 PRINTTAB(9,4):INT(N)
240 Y=460+5*N
250 PLOT 69,F,Y
260 I=INKEY(100*S)
270 IF F›1080 THEN PROCagain
280 F=F+10
290 UNTIL FALSE
300 DEFPROCagain
310 PRINTTAB(0,25);"PLOT AGAIN?YES OR NO"
320 A$=GET$
330 IF A$ ="Y" THEN 30
340 IF A$="N" THEN PRINTTAB(0,25);"PLOTTING IS FINISHED"
```

# 29.6 *Project Module* C5

## Bipolar Transistors

### *What it does*

This Project Module enables high-power devices such as lamps and motors to be switched by low-power devices such as op amps and digital circuits. The bipolar transistor has one npn and one pnp transistor which can be used singly or as a complementary pair. If both transistors are used with the Comparator (Project Module C6), it is possible to control the forward and reverse motion of a d.c. motor. This application is the basis of a simple vehicle which follows a white line marked on a floor. It can also be used to make a solar panel keep track of the Sun.

### *Circuit*

Figure 29.19 shows that this Project Module comprises an npn transistor, $Tr_1$, and a pnp transistor, $Tr_2$, arranged so that they may be used individually or as a complementary pair. These three options are selected by means of the three-position switch, $SW_1$. Resistors $R_1$ and $R_2$ can be selected as collector load resistors for some applications of the transistors, and $R_3$ and $R_4$ provide protection from excess current flow into the bases of the transistors.

### *Components and materials*

$Tr_1$: npn transistor, type TIP31A
$Tr_2$: pnp transistor, type TIP32A
$R_1$ to $R_4$: fixed-value resistors, values
    1 kΩ, 1 kΩ, 220 Ω, and 220 Ω,
    respectively, 0.25 W ± 5%
$SW_1$: miniature, three-position, double-
    pole slider switch
battery: PP9
battery clip: PP9 type
wire: multistrand, e.g. 7 × 0.2 mm
PCB: 90 mm × 50 mm
connectors: PCB header and PCB socket
    housing; crimp terminals

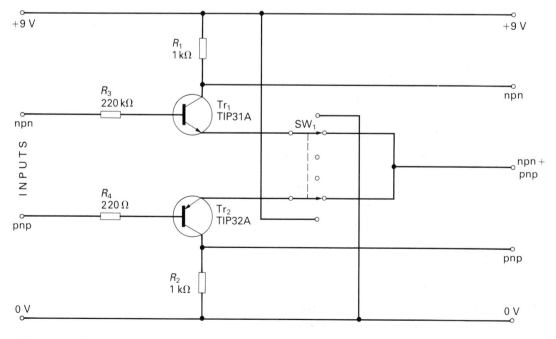

**Figure 29.19** Circuit diagram of the Bipolar Transistors

## PCB assembly

Figure 29.20 shows the layout of the components on the PCB, and figure 29.21 the copper track pattern on the other side of the PCB. See Book A, Section 6.3, for guidance on the preparation of the PCB.

The terminal leads on switch $SW_1$ need trimming before they can enter the holes on the PCB. Note that the transistors do not need heatsinks unless this Project Module is called upon to operate d.c. motors which draw more than 3 A from the power supply.

The terminal pins for the connections on the PCB are made by cutting single and double pins from the PCB header. Single sections of the PCB socket housing are used for terminating wire ends for connecting the bipolar transistors to the

battery and to other Project Modules. Strip 5 mm of insulation from the ends of the wires, and use a crimping tool to squeeze a crimp connector on the bare ends. Push the crimp connector into the PCB socket until it clicks into place.

## Testing and use

Switch $SW_1$ to the 'npn' position. Connect a small d.c. motor, e.g. a Meccano motor, across the 'npn' output (i.e. to the collector) of $Tr_1$ and the positive power supply connection, i.e. the motor is connected in parallel with $R_1$. Connect an LDR across the 'npn' input connection and the positive power supply connection. Change the illumination of the LDR and the motor will switch on and off.

Similarly, check the operation of the pnp

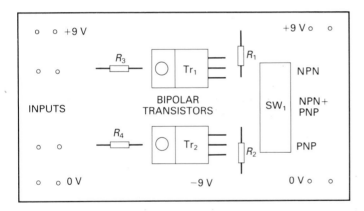

**Figure 29.20** Component layout on the PCB (actual size)

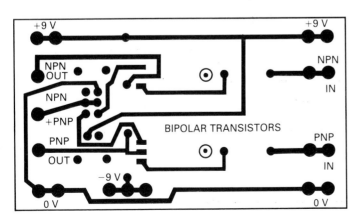

**Figure 29.21** Track pattern of the PCB (actual size)

transistor as follows. Switch $SW_1$ to the 'pnp' position. Connect the d.c. motor across the 'pnp' output connection and the 0 V supply connection, i.e. in parallel with $R_2$. Connect the LDR across the 'pnp' input connection and the 0 V power

supply connection. Change the illumination of the LDR to switch the d.c. motor on and off. The use of the npn and pnp transistors as a complementary pair requires the use of Project Module C6 (Comparator).

# 29.7 *Project Module* C6

## Comparator

### *What it does*

This Project Module compares the voltages of two input signals and produces an output signal when the two voltages are not of equal magnitude. Comparators are used extensively in circuit design, and the comparator can be used with other Project Modules to design a thermostat for the automatic control of temperature, and as the basis of a simple servosystem for the forward and reverse control of a d.c. motor in, for example, a solar panel that automatically tracks the Sun.

### *Circuit*

The circuit shown in figure 29.22 is based on an integrated circuit operational

amplifier (op amp), $IC_1$. The op amp has two inputs, pins 2 and 3, which compare the voltages of the two input signals. At the output, pin 6, a high or low signal is produced depending on which input is at the higher voltage. The resistor, $R_1$, provides the comparator with a small amount of positive feedback so that it operates as a 'snap action' switch. The use of op amps as comparators is discussed in Book D, Chapter 4.

The op amp can be operated from a single power supply of 0 V (pin 4) and + 9 V (pin 7), or from a dual, or 'split' power supply of − 9 V (pin 4), 0 V, and + 9 V (pin 7). The choice of single or dual power supply depends on the application, and an example of each type of supply is described below. The power supply can be from one or two 9 V batteries or from a mains-operated power supply. Chapter 9 describes the design of single and dual mains-operated power supplies. The op amp power supply voltage can be in the range 4.5 V to 15 V, though 9 V batteries are used in the applications below.

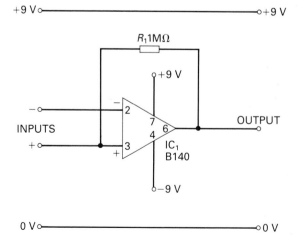

### *Components and materials*

$IC_1$: operational amplifier, 741 or 3140 type
IC holder: 8-way for $IC_1$
$R_1$: fixed-value resistor, 0.25 W, ± 5%
terminal block: 5-way length
battery: 9 V PP9, 2 off, or dual power
    supply
battery clip: PP9 type, 2 off

**Figure 29.22** Circuit diagram of the Comparator

wire: multistrand, e.g. 7 × 0.2 mm; single
strand: e.g. 1 × 0.6 mm
1 × 0.6 mm, for wire links between the
terminal block and the PCB
PCB: 90 mm × 50 mm
connectors: PCB header and PCB socket
housing; crimp terminals

## PCB assembly

Figure 29.23 shows the layout of the
components on the PCB, and figure 29.24
the copper track pattern on the other side
of the PCB. See Book A, Section 6.3, for
guidance on the preparation of the PCB.
Use 6BA nylon screws to secure the
terminal block to the PCB. The terminal
pins for the connections on the PCB are
made by cutting single and double pins

from the PCB header. Single sections of
the PCB socket housing are used for
terminating wire ends for connecting the
Comparator to the battery and to other
Project Modules. Strip 5 mm of insulation
from the ends of the wires, and use a
crimping tool to squeeze a crimp
connector on the bare ends. Push the
crimp connector into the PCB socket until
it clicks into place.

## Testing and use

Use the Comparator and the Relay Driver
(Project Module B2) to wire up the control
system shown overleaf in figure 29.25.
Make sure $IC_1$ is a 3140 op amp. Adjust
$VR_1$ so that when $LDR_1$ is covered and

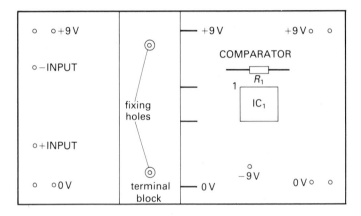

**Figure 29.23** Component layout on the PCB (actual size)

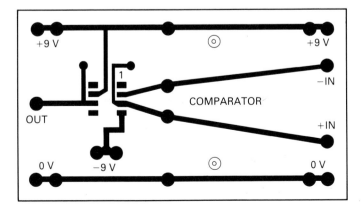

**Figure 29.24** Track pattern on the PCB (actual size)

uncovered the relay switches on and off. The relay contacts can be used to switch power on and off to a d.c. motor or lamp, and it therefore forms the basis of a simple light-operated control system. Book D, Chapters 3 and 4 describes uses of the op amp as a comparator.

# Questions

1  Use a thermistor in place of the LDR and design a thermostat to control the temperature inside a small enclosure using a 12 V, 2.2 W lamp as a heater.
2  Use the comparator and the bipolar transistors to wire up the control system shown in figure 29.26.

Wire the npn and the pnp inputs on the Bipolar Transistor together and connect them to the output of the Comparator. Set $SW_1$ on the bipolar transistors to its centre position. Use a dual power supply of + 9 V and connect the − 9 V terminal of the supply to the − 9 V pin on the Module. Wire the npn output pin to the + 9 V supply and the pnp output pin to the − 9 V supply. Connect a low-power d.c. motor between the npn/pnp pin and 0 V. Vary the relative illumination of the LDRs to control the direction of rotation of the motor.

How could you use this design to make a vehicle follow a source of light? Or a solar panel follow the Sun? See Book D, Chapter 11 for applications of this circuit.

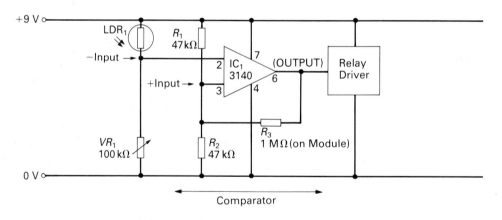

**Figure 29.25**  Light-operated control system

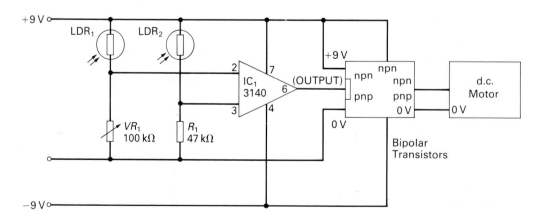

**Figure 29.26**  d.c. motor reversing circuit

# 29.8 *Project Module* C7

## Strain Meter

### *What it does*

This Project Module shows when an object, such as a strut on a crane, changes its shape. The strain is sensed by a strain gauge which is glued to the object being stressed. The change in resistance of the strain gauge produces a change of reading of the meter. An analogue or digital meter can be used which has an f.s.d. of 1 V d.c., e.g. the Voltmeter (Project Module A4). The deflection of the meter shows whether the strain is compressive, i.e. a reduction in length, or tensile, i.e. an increase in length. The Strain Meter is not intended to be connected to other Project Modules, but can be used with the Comparator (Project Module C6) in alarm and control systems.

### *Circuit*

Figure 29.27 shows how a single op amp, $IC_1$, is used to amplify the output from a network of four resistors, $R_1$ to $R_3$, and the strain gauge, $SG_1$. Since the nominal resistance of the strain gauge recommended for this circuit is 120 ohms,

resistors $R_1$ to $R_3$ should have a value of 120 ohms. These resistors should have a tolerance of better than 0.1% if temperature changes are not to cause unwanted changes in the reading of the meter. To cancel the effects of temperature on the strain gauge, a second strain gauge should replace resistor $R_3$. This second strain gauge should be mounted alongside the first strain gauge on the specimen being strained, but not allowed to experience the same strain. However, the circuit shown gives serviceable results without the use of a second temperature-compensating strain gauge. Book B, Chapter 7, discusses the properties of strain gauges.

When the strain gauge is unstrained, variable resistors $VR_1$ (coarse control) and $VR_2$ (fine control) are set to give a 'zero' reading on the meter, M. This 'zero' voltage may be any convenient value, e.g. 1.5 V, and not necessarily equal to the 0 V supply for the circuit. An increase in the reading on the meter means that the strain gauge is undergoing an increase in length (tensile strain). If the meter reading decreases, the strain gauge must be

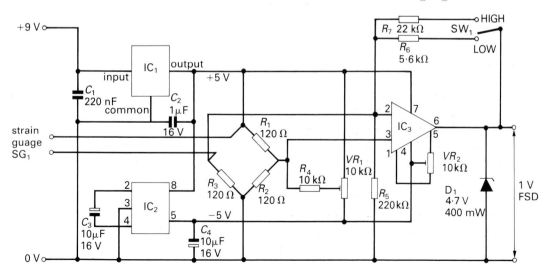

**Figure 29.27** Circuit diagram of the Strain Meter

undergoing a decrease in length (compressive strain).

The variable resistor, $VR_3$, adjusts the sensitivity of the op amp, $IC_3$, and therefore enables the circuit to respond to different magnitudes of strain. Stable readings on the meter are assured by the use of $+5$ V and $-5$ V supply voltages by means of the voltage regulators $IC_1$ and $IC_2$, respectively. The use of an op amp to amplify the output from a Wheatstone bridge is described in Book D, Chapter 6.

## Components and materials

$IC_1$: 5 V, 1 A voltage regulator, type 7805
$IC_2$: $+5$ V to $-5$ V voltage regulator, type 7660
$IC_3$: operational amplifier, type 741
$D_1$: Zener diode, type 4.7 V, 400 mW
IC holder: $2 \times 8$-way

$SG_1$: foil strain gauge, nominal resistance 120 $\Omega$
$C_1$: polyester capacitor, value 220 nF
$C_2$ to $C_4$: electrolytic capacitors, values 1 $\mu$F, 10 $\mu$F and 10 $\mu$F, respectively, 16 V working
$VR_1$: 10 k$\Omega$ trimmer variable resistor
$VR$: 19.1 mm long 20-turn variable resistor, 10 k$\Omega$
$R_1$ to $R_3$: fixed-value resistors, 0.33 W, $\pm 0.1\%$, or, if reduced circuit stability can be tolerated, 0.4 W, $\pm 1\%$
$R_4$ to $R_7$: fixed-value variable resistors, values 10 k$\Omega$, 220 k$\Omega$, 10 k$\Omega$, 5.6k$\Omega$ and 22 k$\Omega$, 0.25 W, $\pm 5\%$
$SW_1$: single pole, change-over slide switch
battery: PP9
battery clip: PP9 type
wire: multistrand, e.g. $7 \times 0.2$ mm; single strand, e.g. $1 \times 0.6$ mm
PCB: 90 mm $\times$ 50 mm

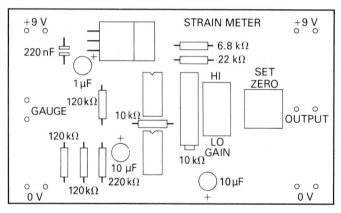

**Figure 29.28** Component layout on the PCB (actual size)

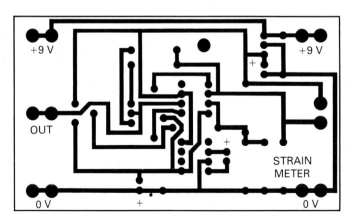

**Figure 29.29** Track pattern on the PCB (actual size)

connectors: PCB header and PCB socket housing; crimp terminals

## PCB assembly

Figure 29.28 shows the layout of the components on the PCB, and figure 29.29 the copper track pattern on the other side of the PCB. See Book A, Section 6.3, for guidance on the preparation of the PCB.

The terminal pins for the connections on the PCB are made by cutting single and double pins from the PCB header. Single sections of the PCB socket housing are used for terminating wire ends for connecting the Strain Meter to the battery, to the meter and to the strain gauge. Strip 5 mm of insulation from the ends of the wires, and use a crimping tool to squeeze a crimp connector on the bare ends. Push the crimp connector into the PCB socket until it clicks into place.

Use Araldite to glue the strain gauge to the surface of a 150 mm × 10 mm × 1 mm strip of plastic. Solder fine insulated leads to the solder pads on the strain gauge. Tape the fine leads to the steel strip, then solder more robust leads to these fine leads and terminate them with PCB connector sockets so that the strain gauge can be connected to the Strain Meter. Do not use the strain gauge for 24 hours.

## Testing and use

Plug the Strain Gauge into the Strain Meter. Connect a d.c. voltmeter, e.g. the Voltmeter (Project Module A4), to the output of the strain meter and a PP9 battery to the power supply connections. Use a digital multimeter if you have one so that negative voltages can be measured since this makes setting up easier.

The Strain Meter needs setting up so that, when the strain gauge is unstrained, the output voltage is about 0.5 V. If the steel strip is then bent so that the strain gauge gets slightly longer, the output voltage should increase (tensile stress); and if it is bent so that the strain gauge gets slightly shorter (compressive stress), the output reading should decrease. The setting up is done by adjusting the three multiturn trimmers, $VR_1$ to $VR_3$, using an adjusting tool made for the job. Proceed as follows:

**(a)** Set the multimeter to its 5 V d.c. (or nearest) scale.
**(b)** Set $SW_1$, to the 'LOW' position.
**(c)** Adjust $VR_1$, the 'coarse' zero control until the reading on the voltmeter is about + 0.5 V. You may need to adjust $VR_2$, the 'fine' zero control in order to achieve this.
**(d)** Check that the reading on the voltmeter changes when you bend the plastic strip.
**(e)** Now switch the multimeter to its 1 V range. Bend the strip again and the response will be more marked. If the response is not sensitive enough, switch $SW_1$ to its 'HIGH' position. Readjust $VR_1$, if necessary to make sure the reading is still on the scale. The gain of the circuit may be increased further by increasing the values of $R_6$ or $R_7$.

The Strain Meter can now be used for examining compressive and tensile forces in members of a structure, or it can be used in pressure-sensitive alarms, overload alarms, etc.

## Questions

1  Use the Comparator (Project Module C5) with the Strain Meter so that an alarm sounds when the plastic strip is bent a certain amount. You will also need the Pulser (Project Module A2) and the Piezo-sounder (Project Module A3).
2  How can you use the Comparator (Project Module C6) with a dual voltage power supply so that a d.c. motor is off when the strain gauge is unstrained, but changes direction as the plastic strip is bent first one way then the other way?

**3** Design a pressure-operated door opening and closing system. When a person steps on a pressure pad (the strain gauge on the plastic strip), the door opens; the door closes when the person steps off the pad.

If your computer is fitted with an analogue-to-digital converter (ADC) (Chapter 10, Book D), the readings from the Strain Meter can be displayed on a VDU. The two Basic programs below are written for the BBC microcomputer which has an analogue port for reading changing voltages. A suitable connector unit that plugs into this port can be obtained from Technology Teaching Systems Ltd, Penmore House, Hasland Road, Hasland, Chesterfield, S41 0SJ. This unit is supplied in kit form. The four analogue channels of the BBC microcomputer are accessible through four 4 mm sockets on the connector unit. The BBC's 1.8 V reference and 5 V supply voltages are also available on this unit.

## Program 1 Max/Min Strain Meter

This simple program displays the current strain (in arbitrary units) and the maximum and minimum strains reached in a period of time.

Lines 40 to 70 display the three temperatures.
Line 80 assigns a variable to the output voltage from the Strain Meter which is accessed via channel 1 of the analogue port. Division by 360 converts the analogue reading to a number which represents the 'units' of strain.
Line 120 initially prints the value '0' at each of the strain positions.
Line 130 introduces a delay of 1 second between readings.

Lines 140 to 160 read the analogue input voltage again and produce a variable (V) which is the difference betwen consecutive readings.
Line 170 checks if the readings have changed; if not the program loops back to continue displaying the same reading.
The procedure at line 180 decides whether to update the maximum strain or the minimum strain; if the current strain has increased since the last reading, the value of Q is updated and printed at line 120; if the current strain has decreased since the last reading, the value of R is updated at line 120.

## Program 2 Strain/Time Plot

This program produces a graph of strain against time.

Line 20 selects a graphics mode that displays large-size printing on the VDU.
Lines 30, 40 and 50 identify three procedures that structure the program.
The procedure at line 60 designs a graphics window in which the graph is plotted and labels the axes of the graph.
The procedure at line 130 draws in the axes which allow positive and negative strain to be displayed.
The procedure at line 160 plots the graph after the user has decided (lines 170, 180 and 300) the interval between readings.
Line 140 prints the value of the current strain in the graphics window.
Lines 250 and 260 select the reading from channel 1 of the analogue port and convert it to a reading representing strain.
Line 270 provides a variable which is the difference between consecutive readings.
Lines 280 and 290 plot a graph which automatically sets the unstrained output from the strain meter to '0'.
The procedure at line 360 asks if readings are to be repeated.

## Program 1

```
 10 REM "MAX/MIN STRAIN METER"
 20 MODE5
 30 CLS
 40 PRINTTAB(0,6);"MAX/MIN STRAIN METER"
 50 PRINTTAB(0,12);"NOW STRAIN=      UNITS"
 60 PRINTTAB(0,14);"MAX STRAIN=      UNITS"
 70 PRINTTAB(0,16);"MIN STRAIN=     UNITS"
 80 CH0=ADVAL(1)
 90 A=INT(CH0/360)
100 P=0:R=0:Q=0
110 REPEAT
120 PRINTTAB(11,12);P;TAB(11,14);Q;TAB(11,16);R
130 I=INKEY(100)
140 CH0=ADVAL(1)
150 X=INT (CH0/360)
160 V=X−A
170 IF V◊P THEN PROCmaxmin ELSE 120
180 DEFPROCmaxmin
190 IF V›Q THEN Q=V ELSE Q=Q
200 IF V‹R THEN R=V ELSE R=R
210 IF P‹10 THEN PRINTTAB(12,12);"    "
220 IF P>99 THEN PRINTTAB(13,12);"   "
230 P=V
240 UNTIL FALSE
```

## Program 2

```
 10 REM"STRAIN/TIME PLOT"
 20 MODE5
 30 PROCwindow
 40 PROCaxes
 50 PROCplot
 60 DEFPROCwindow
 70 CLS
 80 VDU 24,30;400;1179;1000;:GCOL0, 130
 90 CLG
100 VDU5:GCOL0,0:MOVE55,940:PRINT"+":MOVE55,880:PRINT"s":MOVE55,
    820:PRINT"t":MOVE 55,760:PRINT"r":MOVE55,700:PRINT"a":MOVE55,640:PRINT"I"
    MOVE55,580:PRINT"n":MOVE55,520:PRINT"−"
110 VDU5:MOVE1080,680:PRINT"t" :GCOL0,1:MOVE350,950:PRINT"STRAIN/TIME":VI
120 ENDPROC
130 DEFPROCaxes
```

```
140  GCOL0,0:MOVE 120,710:DRAW1080,710:MOVE120,460:DRAW120,960
150  ENDPROC
160  DEFPROCplot
170  PRINTTAB(0,25);"PLOT RATE?:ENTER SEC"
180  INPUT S
190  F=130
200  CH0=ADVAL(1)
210  A=INT(CH0/360)
220  P=0
230  REPEAT
240  PRINTTAB(9,4);P
250  CH0=ADVAL(1)
260  X=INT(CH0/360)
270  V=X−A
280  Y=710+5*V
290  PLOT 69,F,Y
300  I=INKEY(100*S)
310  IF F>1080 THEN PROCagain
320  IF V<10 OR V>−10 THEN PRINTTAB(10,4);"        "
330  F=F+10
340  P=V
350  UNTIL FALSE
360  DEFPROCagain
370  PRINTTAB(0,25);"PLOT AGAIN?YES OR NO"
380  A$=GET$
390  IF A$="Y" THEN 30
400  IF A$="N" THEN PRINTTAB(0,25);"PLOTTING IS FINISHED"
```

The Zener diode, $D_1$ in figure 29.28 ensures that the BBC microcomputer cannot be damaged by an excessive voltage fed to its analogue port. Note also that the Strain Meter should be operated from its own 9 V d.c. supply — a 9 V battery is very suitable. The output of the Strain Meter should be connected to channel 0, i.e. ADVAL(1), of the connector unit, and the 0 V supply connected to the 0 V plug on the connector unit.

# Questions and answers — Book C

## Revision questions

If you have taken the quicker route through this book, do not answer questions marked with an asterisk(*).

### *Diodes*

1 Diodes are used to store electric charge in circuits.
   True or false?

2 Diodes are unlike ordinary resistors because they allow current to flow through them easily in only one direction.
   True or false?

3 If a voltage is applied across a diode so that current easily flows through it, the diode is said to be . . .
   forward-biased/reverse-biased/ unstable?

4 As the voltage across a junction diode increases in the forward-biased direction, the current . . .
   increases/decreases/reverses?

5 How should a battery be connected to a diode to make a large current flow through it? What happens if the battery is connected the other way round?

6 Which lamp lights in the circuit shown in figure 30.1?

7 State two differences between the forward-biased and reverse-biased regions of the diode characteristic.

8 As the reverse-bias voltage across a diode increases, very little current flows until the . . . . voltage is reached.
   breakdown/maximum/short-circuit?

9 At the breakdown voltage, a . . . . current flows.
   small/negligible/large?

10 Rectification is the process of converting a direct voltage into an alternating voltage.
   True or false?

11 'Half-wave rectification' means that half of one cycle of an alternating voltage is removed by a diode.
   True or false?

12 Which lamp lights in the circuit shown in figure 30.2?

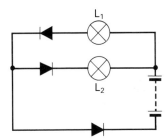

**Figure 30.1**

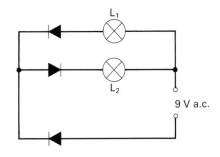

**Figure 30.2**

13 Capacitors are used in rectifier circuits to . . . . a pulsing direct voltage.
    rectify/smooth/cut out?

14 What is meant by the 'ripple voltage' in a rectifier circuit?

15 In general, as the load across the output of a rectifier circuit increases, the ripple voltage increases.
    True or false?

16 Name two other uses of diodes besides rectification.

17 The first diagram in figure 30.3 shows another use for a diode. The delicate movement of a moving-coil meter is protected by the series diode against wrongly connecting the leads to a d.c. supply, when measuring current for example. Only if the diode is forward-biased does the meter register a current. The wrong connection would reverse-bias the diode. The same principle can be used to protect a circuit against the error of wrong battery connections, as the second diagram shows. How does this protection work?

18 Another way of protecting a meter is shown in the third diagram in figure 30.3. How does this work?

## Zener diodes

19 The Zener diode is a diode which has a very . . . . resistance up to the reverse breakdown voltage.
    high/low/unequal?

20 At the breakdown voltage for a Zener diode, the reverse resistance of the diode increases to a very high value.
    True or false?

21 The Zener diode owes its usefulness to the fact that, at and above the breakdown voltage, large changes of current are produced by . . . . changes of voltage.
    small/large/equal?

22 Zener diodes are used in circuits where the output voltage from a source needs to be stabilised.
    True or false?

## Resistivity and conductivity

*23 An element which is widely used in the manufacture of junction diodes is . . . .
    carbon/silicon/metal?

*24 Silicon and germanium both have four electrons in their outer shell.
    True or false?

*25 The electrons in the outer shell of an atom are known as . . . . electrons.
    fast/valence/spare?

*26 Free electrons must be available in a material if it is to conduct electricity easily.
    True or false?

*27 What effect has heat on the conductivity of a piece of silicon?

*28 The conductivity of silicon may be increased by the addition of carefully chosen impurities.
    True or false?

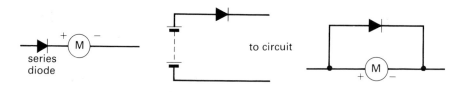

**Figure 30.3**

## Semiconductors

**\*29**  A p-type semiconductor could be made by adding . . . . atoms to silicon.
water/copper/boron?

**\*30**  A pentavalent impurity atom produces an excess of unshared electrons when it is introduced into a piece of germanium.
True or false?

**\*31**  The charge carriers in n-type material are mostly electrons. These are called . . . . carriers in n-type material.
majority/minority/heavy?

**\*32**  When a junction diode is forward-biased, electrons and holes flow across the junction easily.
True or false?

**\*33**  When a junction diode is reverse-biased, the depletion region is widened.
True or false?

## Bipolar ransistors

**34**  Since the transistor can produce power amplification, it is known as an 'active' component.
True or false?

**35**  Name the three parts to the structure of bipolar transistors.

**36**  The leads on germanium transistors are generally . . . . than on silicon transistors.
shorter/longer?

**37**  The transistor shown in figure 30.4 is a . . . . type.
pnp/npn?

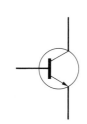

**Figure 30.4**

**\*38**  Germanium and silicon are semiconductors. This means that they . . .
are heavy elements/withstand high and low temperatures well/have an electrical conductivity midway between that of metals and insulators?

**39**  One advantage of silicon transistors, compared with germanium transistors, is that they are more tolerant of being accidentally overheated.
True or false?

**40**  The symbol for an npn transistor can always be distinguished from that of a pnp transistor because . . .
the npn symbol is larger than the pnp symbol/the arrow in the pnp symbol points towards the base lead/the pnp symbol is inverted?

**41**  Most modern silicon transistors are npn types.
True or false?

**\*42**  The silicon planar construction is favoured for all modern transistors because it is more useful than the alloy-diffused type at . . . . frequencies.
high/low?

**\*43**  It is important for the base region of a transistor to be . . .
thin/thick/elastic?

**\*44**  The base region of an npn transistor is made of . . . . semiconducting material.
p-type/n-type/thick?

**\*45**  N-type semiconducting material is made by doping silicon or germanium with . . .
neutral atoms/arsenic or phosphorus atoms/indium or aluminium atoms?

**\*46**  An acceptor impurity atom is an atom which accepts holes.
True or false?

**\*47** The base region of a transistor is . . . . doped with impurity atoms compared with the other two regions.
        lightly/heavily?

**48** An npn or pnp transistor has two p-n junctions.
        True or false?

**49** Which junction of a transistor is forward-biased when the transistor is connected in the common-emitter mode?

**50** The base-collector junction of a transistor is normally reverse-biased. This is to ensure that . . .
        a large collector current flows across this junction/the transistor does not become too warm/almost zero current flows across this junction?

**\*51** There is a very small current across a reverse-biased p-n junction. This is known as the . . .
        leakage current/surprise current/ forward current?

**\*52** The majority carriers are . . . . in p-type semiconducting material.
        holes/electrons?

**\*53** When a transistor is biased normally, the base current is always greater than the emitter current.
        True or false?

**\*54** The purpose of making the base region thin and lightly doped with impurity atoms is to allow . . . . of the majority carriers to reach the collector region from the emitter region.
        as many as possible/none/as few as possible?

**\*55** Transistor action is the turning-on of a current through the . . . . resistance base-collector junction by a current passing through the . . . . resistance of the base-emitter junction.
        high/low?     high/low?

**56** Since a small base current can switch on a large collector current, the transistor acts as a . . .
        transistor amplifier/current amplifier/resistance amplifier?

**57** The current flowing into a transistor . . . . equals the current flowing out of it.
        always/never/sometimes?

**58** Under certain circuit conditions, the npn transistor type 2N3053 requires a base current of 4 mA to produce a collector current of 340 mA. What is the d.c. current gain of the transistor under these conditions?

**59** Figure 30.5 shows an npn transistor operating under circuit conditions which enable it to produce a d.c. current gain of 75. Draw the circuit and mark in the currents which flow.

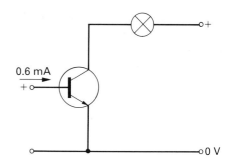

**Figure 30.5**

**60** The current flow across the forward-biased p-n junction of a silicon transistor is only significant when the voltage across this junction just exceeds . . .
        0.2 V/0.6 V/0.9 V?

**61** The switch-on voltage for a p-n junction is the voltage required to . . .
        cause a large current to flow across it/reverse the current through it/heat up the junction?

**\*62** What do the output characteristics of a transistor tell you?

**\*63** Figure 30.6 shows some typical output characteristics of a transistor. Estimate the d.c. current gain of this transistor when it is operated at the four points shown.

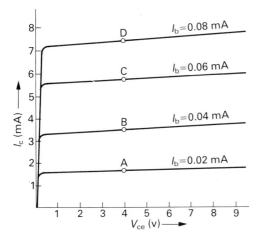

**Figure 30.6**

**\*64** The circuit diagram in figure 30.7 applies to the output curves also of question 63. Suppose the base bias resistor $R_b$ is small enough to make the transistor draw the largest current from the supply. What is the collector current $I_c$? Mark this point on the curves with the letter X.

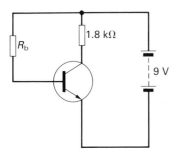

**Figure 30.7**

**\*65** What current is drawn by the transistor when the transistor is cut-off? Mark this point on the curves in the figure with the letter Y.

**\*66** Draw a line between the points X and Y. What is the name given to it?

**\*67** From this line, estimate the base current when $I_c = 2.5$ mA. What is the d.c. current gain at this point?

**\*68** Calculate the value of $R_b$ if the base-emitter voltage is to be 0.7 V.

**\*69** When a transistor is operating under saturation conditions, the emitter-collector voltage is almost zero.
True or false?

**70** A transistor is being tested, and it is found that the voltage across the collector-emitter junction is zero when there is no base current. Is this transistor a dud?

**71** Electrical power dissipation in a transistor produces . . .
electricity/heat/resistance?

**\*72** Electrical power = amperes × volts =
. . .
resistance/watts/heat?

**\*73** Watts = (. . . .)$^2$ × ohms.
resistance/volts/amperes?

**\*74** When a transistor is operating under saturation conditions, the electrical power dissipated in it is high.
True or false?

## Field-effect transistors

**75** Draw the circuit of a simple touch switch using a field-effect transistor.

**76** A field-effect transistor has a . . . . input resistance compared with a bipolar transistor.
high/low?

**77** Explain how the junction-gate field-effect transistor acts as a voltage-controlled resistor.

**78** What does 'unipolar' mean in relation to an FET?

**79** What does CMOS stand for?

**80** Explain the structure of the VMOS field-effect transistor.

# Revision answers

**58** 85
**60** 0.6 V
**63** A — 90; B — 97; C — 100; D — 96.
**64** 5 mA
**65** 0 A
**67** 30 μA; 83
**68** 270 kΩ (approximately)

# Answers to questions

Section 3.1
  **1** Preferred value 270 Ω
Section 5.3
  **1** $\frac{1}{50}$
Section 13.2
  **1** 14

Section 18.2
  **1** 0.1 mA
  **2** 0.25 mA; 49.75 mA
Section 19.1
  **1** 34; 185 mA
  **2** 400
Section 20.4
  **1** 36 000
Section 21.5
  **1** Preferred value 430 kΩ
  **2** 120 mW
Section 23.2
  **1** 50
  **2** 0.6 mA; 100
Section 25.4
  **2** 100
  **3** 4 mA